Mechanical Properties of Bamboo

FORESTRY SCIENCES

Volume 37

The titles published in this series are listed at the end of this volume.

Mechanical Properties of Bamboo

JULES J.A. JANSSEN
Eindhoven University of Technology
Faculty of Architecture and Building
Eindhoven, The Netherlands

Kluwer Academic Publishers
Dordrecht / Boston / London

Library of Congress Cataloging-in-Publication Data

Janssen, Jules J. A.
Mechanical properties of bamboo / by Jules J A. Janssen.
p. cm. -- (Forestry sciences ; v. 37)
Includes indexes.
ISBN 0-7923-1260-0 (HB : acid free)
1. Bamboo--Mechanical properties. I. Title. II. Series.
TA455.B3J36 1991
620.1'2--dc20 91-14533

ISBN 0-7923-1260-0

Published by Kluwer Academic Publishers,
P.O. Box 17, 3300 AA Dordrecht, The Netherlands.

Kluwer Academic Publishers incorporates
the publishing programmes of
D. Reidel, Martinus Nijhoff, Dr W. Junk and MTP Press.

Sold and distributed in the U.S.A. and Canada
by Kluwer Academic Publishers,
101 Philip Drive, Norwell, MA 02061, U.S.A.

In all other countries, sold and distributed
by Kluwer Academic Publishers Group,
P.O. Box 322, 3300 AH Dordrecht, The Netherlands.

Printed on acid-free paper

Printed in the Netherlands

Contents

Preface

In this book I have collected many reports on the properties of bamboo. Since I started bamboo research in 1974, I have collected many publications on bamboo. However, in contacts with other researchers I became aware of the fact that several of these publications were unknown to them. Consequently their activities in research or in bamboo projects had to start at a lower level than if they had known some of these publications. Therefore, I discussed the need for a book like this with several colleagues, and I decided to start writing.

In this book I present data on properties, as published by researchers. Certainly it is not complete; e.g. the proceedings of the bamboo workshop in Cochin in November 1988 are not included; they arrived just too late to be taken into account. On the other hand, in writing this preface I realize I have forgotten my own thesis. I do apologize to any colleague whose relevant report is not taken into account as well.

I like to express my feelings of respect to the board of my faculty: they allowed me to spend so much time on this book, and they gave a considerable financial support for the printing costs.

In my opinion bamboo should become a completely normal building material, just as timber, steel or concrete. I realize it is still a long way to go to reach this purpose. I hope this book is a small step on that way. I would appreciate if the reader is so kind to send me his comments.

Eindhoven, January 1991
Jules J.A. Janssen

1. Growth and anatomy

GROWTH 1

Keyword: Form of the culm.
Subject: The internode-length, diameter and wallthickness along the length of the culm.
Source: Shigematsu, Y. Analytical investigation of the stem form of the important species of bamboo. Bull. Fac. Agr. Univ. Miyazaki, vol 3 pp 124-135 (1958). Libr.: Min. of Agr., London.

Results: The author has studied many culms of 15 different bamboo species from Miyazaki in Japan.

At first, he has studied the length of the internode as a function of the internode-number. This increases from the bottom part of the culm towards a point somewhere below the middle, and then decreases till the top of the culm. The relation is given by the formula:

$y = a + b*x + c*x^2$,

in which:

y = the length of the internode, given as a percentage of the maximum length which is 100%.

x = the number of the internode, given as a percentage: 0% is bottom, 100% is top.

In the lower part of the culm this formula is: (mean value)

$y = 25.13 + 4.8080*x - 0.0774*x^2$, and in the upper part:

$y = 178.84 - 2.3927*x + 0.0068*x^2$.

The original paper gives these formulas for all 15 species.

In the original paper also 15 plots are drawn with (for each species) the internode-length, the diameter, the wallthickness and the wallthickness-index, all plotted against the internode-number (the wallthickness-index is: 2*wallthickness/diameter, multiplied by 100).

Similarly, the author gives information on the diameter as function of the internode-number. From the detailed information in the

original paper we quote only the mean value for all 15 species:
for the lower part: $y = 97.5 - 0.212*x - 0.016*x^2$,
(the original paper gives + 0.016, but that is a misprinting),
for the upper part: $y = 157.6 - 2.868*x + 0.013*x^2$, or
for the whole culm: y = 101.8*e^(-0.022*x), in which

y = the diameter, 100% at the bottom, decreasing towards 0% at the top.

x = the internodenumber, 0% at the bottom and 100% at the top.

For the wallthickness-index the formula is (mean for all 15 species):

y = 35 + 0.0181*(x - 35)^1.90, in which

y = wallthickness index and x = the internodenumber, both in percentages; y varies from a 50% at the bottom to a 35% below the middle and then to a 80% at the top.

GROWTH 2

Keyword: Form of the culm.
Subject: The distribution of the sizes of internodes along the length of a culm, etc.
Source: Fangchun, Z. Studies on the structure of culm form of Phyll.pub.; J. of Nanjing techn. coll. of for. prod., 1981 no 1, pp 16-69, in chinese with short english summary. Also published under the title: The structure of culm form of Phyll.pub., in: Bamboo Research, edited by W.Y. Hsiung, 1981 vol 1, Nanjing techn. coll. of for. prod, pp 75-98, in English.

Results: The researcher Zhou Fangchun has studied 3567 culms, collected from 13 locations in China and Japan, from the species Phyll.pub. He has derived mathematical formulae for the following relations:

1. The distribution of the internode lengths along the culm, related to the node-number.
2. Idem for the internode circumference.
3. Idem for the internode culmwall thickness.
4. Idem for the internode volume.
5. The relative diameter of the culm, related to the height.
6. Idem for the relative culmwall thickness.
7. Idem for the relative fresh weight.
8. The relation between the culm breast-height diameter and the culm length, which is dependent on the annual mean temperature and rainfall.
9. Idem for the culm breast-height diameter and the fresh weight.
10. Idem for the culm breast-height diameter and the culmlength.
11. Idem for the culm breast height diameter and the total fresh weight of branches and leaves.
12. The relationship between the culm breast-height diameter and the diameter at any height.
13. The relationship between the breast-height diameter and the culmwall thickness at any height.

14. The relation between the total fresh weight of the culm and the breast-height diameter, the length, the annual mean temperature and the rainfall.

After quite some reflection the actual author has decided not to quote all the tables and formulae in full; this would have required several pages, and it covers a very specific field of interest only.

GROWTH 3

Keyword: Biomass, form of the culm.
Subject: The sizes of the culm, and the weight, for Phyll. bamb.
Source: Watanabe, M. and K. Ueda. On the structure of Madake (Phyll. bamb.). The reports of the Fuji bamboo garden, Dec. 1976, no 21, pp 9-26. In Japanese, with a summary in English.

Results: The authors have studied samples from seven stands of Phyll. bamb. from the outskirts of Kyoto, Japan. Their conclusions are as follows.

1. The relation between the diameter at breast height D and the length of the culm H is:

$$H = 3.4815*D^{0.7114},$$

valid for D = 30 to 120 mm and H = 5 to 20 m.

2. The relation between D^2H (square of diameter at breast height multiplied with culm height) and the dry weight Ws of the culm is:

$$Ws = 0.0255*(D^2H)^{0.844},$$

valid for Ws = 0.5 to 20 kg, and D^2H = 1000 to 100 000 mm^2*m.

3. Idem for the dry weight Wt of the above ground shoot including culm, leaves and branches:

$$Wt = 0.034*(D^2H)^{0.8428},$$

valid for Wt = 0.5 to 50 kg and D^2H = 1000 to 100 000 mm^2*m.

4. The relation between the dry weight of the branches Wb and the dry weight of branches and leaves Wbl per culm is:

$$Wb = 0.7339*Wbl^{0.961},$$

valid between 0.1 and 5 kg for each.

5. The relations between the total dry weight of culms in a 1 by 1 m square plot, Ws, and the dry weight of rhizome Wrh or the dry weight of stumps Wst, all in the same square meter, are:

$$Wst = 0.52*Ws, \text{ and } Wrh = 0.8*Ws,$$

valid for Ws = 1 to 10 kg, Wst = 0.5 to 5, and Wrh = 1 to 10 kg.

6. The biomass in 1000 kg/ha is:

culms:	15.7-61.2,
branches	3.8-13.7,
leaves	1.6-6,
above ground shoot	21-81,
rhizome	12.6-49,
stump	8.2-31.8, and
underground biomass	20.8-80.8.

7. The relation between the above ground biomass Y in 1000 kg/ha, and the total area A of the cross sections at breast height in m^2/ha is:

$$Y = 1.063*A^{1.2166},$$

valid for Y = 10 to 100, and A = 10 to 50.

ANATOMY 1

Keyword: Anatomy, cell, fibre, position.
Subject: Dimensions of cells and vascular bundles.
Source: Hsiung, W., S. Qiao and Y. Li. The anatomical structure of culms of Phyll.pub. Mazel ex H. de Lahaie. Acta botanica sinica (Dec. 1980) vol 22 no 4 pp 343-348 and two plates with photographs.

Results: The dimensions of the cells in µm are:

position in culm	bottom				middle				top				mean			
	D	p	a	L	D	p	a	L	D	p	a	L	D	p	a	L
rind (outer skin)																
epidermis		17	23	37		15	21	43		15	23	67		16	23	49
hypodermis		19	15			12	12			13	10			15	12	
cortex	25	30	19	32	19	23	15	46	18	22	14	74	21	25	16	51
vessels, outer	42				44				34				40			
middle	105				118				99				107			
inner	135				135				137				135			
all				537				594				458				530
ground tissue																
outer	41			148	35			156	26			234	34			176
middle	58			247	51			306	39			384	49			321
inner	66			213	60			237	47			358	58			269
mean	55			215	49			233	37			321	47			255
pith ring																
(inner skin)		79	34	16		93	26	20		98	21	?		90	27	20

in which D= diameter, p= periclinal, a= anticlinal, L= length

(periclinal is the horizontal diameter, anticlinal the vertical).

The sizes of the vessels are:

position in culm	bottom			middle			top			mean		
wall portion	out	mid	inn	out	mid	inn	out	mid	inn	out	mid	inn
length												
out.fibre cap	196	202	217	151	132	111	159	75	74	169	166	134
conduct.bundle	77	302	310	134	252	254	114	219	252	108	258	272
inn.fibre cap	317	254	131	251	191	97	186	168	100	251	204	109
width												
left fibre cap	105	112	115	99	80	100	79	72	71	94	91	95
conduct.bundle	116	295	457	120	318	418	114	285	3?6	117	299	424
right fibre cap	106	111	118	88	89	100	77	72	74	96	91	97
length/width	1.8	1.6	0.9	1.7	1.1	0.6	1.7	1.0	0.7	1.7	1.3	0.8

in which out/mid/inn = outer/middle/inner in the wall thickness

ANATOMY 2

Keyword: Fibre.
Subject: The dimensions of fibres in 13 Philippine bamboos.
Source: Tamolang, F.N., a.o. Properties and utilization of Philippine erect bamboos. Pp 189-200 in: Bamboo research in Asia. Proc. workshop Singapore May 1980. Editors Lessard, G. and Chouinard, A. Ottawa, ISBN: 0-88936-267-X.

Results: Philippine researchers have determined the fibre dimensions for the next species.

Species	L	D	l	w	SR	FR	RR
Schizost. lima	1.67	0.022	0.004	0.009	76	18	4.50
Dendr. merril.	2.16	0.014	0.006	0.004	154	43	1.33
Schizost.diff.	2.00	0.021	0.007	0.007	95	33	2.00
Gigantochl.levis	1.80	0.022	0.006	0.008	82	27	2.67
Schizos.lumampao	2.42	0.014	0.006	0.004	173	43	1.33
Gigantoch.aspera	3.78	0.019	0.007	0.006	199	37	1.71
Bamb.arundinacea	1.73	0.022	0.006	0.008	79	27	2.67
Bamb.multiplex	1.36	0.018	0.002	0.008	76	11	8.00
Bamb.vulgaris	2.33	0.017	0.004	0.007	137	23	3.50
Bamb.blumeana	1.95	0.018	0.004	0.007	108	22	3.50
Phyll.nigra	1.86	0.019	0.006	0.0065	98	32	2.17
Bamb.tulda	1.45	0.200	0.005	0.0075	73	25	3.00
B.vulg.striata	1.66	0.021	0.005	0.008	73	24	3.20

in which:
L = length in mm,
D = width in mm,
l = lumen width in mm,
w = cell wall in mm,
SR= slenderness ratio = L/D,
FR= flexibility ratio = 100*l/D,
RR= Runkel ratio = 2w/l.

ANATOMY 3

Keyword: Fibre, position.
Subject: The length and width of fibres, for many species, and their distribution over the wallthickness and the internodes.
Source: Liese, W. and D. Grosser. Untersuchungen zur Variabilitaet der Faserlaenge bei Bambus (Research into the variability of the length of fibres in bamboos). In German. Holzforschung, vol 26 no 6 (Dec.1972), pp 202-211.

Results: The authors have investigated the dimensions of the fibres in many bamboo species. At first they give for 78 species the length and width of the fibres, as published by other authors; an excellent overview, but too long to be reproduced here. The remaining part of this article gives several tables and graphs dealing with the variation of the length and width of the fibres over the wallthickness, and over the internodes from bottom to top.
This report is very detailed, and it contains many data; however in the opinion of the present author is too detailed to be quoted here to the full.

ANATOMY 4

Keyword: Anatomy, fibre, vessel, position.
Subject: The percentages of parenchyma-tissue, fibres and vessels, related to the position along the height of the culm, or to the wallthickness.
Source: Grosser, D. and W. Liese. Verteilung der Leitbuendel und Zellarten in Sproszachsen verschiedener Bambusarten (distribution of vascular bundles and cell-types in the culms of several bamboo species). In German. Holz als Roh- und Werkstoff, vol 32 (1974), pp 473-482.

Results: The authors have studied the following species:

Cephalostachyum pergracile, from Dehra Dun, India,
Phyllostachys edulis, from Taiwan,
Phyllostachys makinoi, from Taiwan,
Schizostachyum brachycladum, f1 3 4ia.

They have studied the percentages of vascular bundles, parenchyma and fibres varying with the position in the wallthickness (outer, middle and inner) and in the internodes from bottom to top. From the complete tables only some mean values will be quoted here.

	Ceph.per.			idem			Phyll.mak			Schiz.br.			idem			Tein.dull		
	Pa	Fi	Va	Pa	Fi	Va	Pa	Fi	Va	Pa	Fi	Va	Pa	Fi	Va	Pa	Fi	Va
o	42	52	5.3	41	54	4.6	36	59	5.3	39	55	5.8	36	59	4.4	37	58	4.9
m	52	39	9.0	49	43	8.2	53	37	10	54	37	8.5	56	35	8.6	51	40	8.9
i	65	26	9.9	62	29	9.6	75	16	8.6	66	22	11	69	19	12	67	22	11
M	53	39	8.1	50	42	7.5	55	37	8.0	53	38	8.6	54	38	8.3	52	40	8.2

in which Pa = parenchyma, Fi = fibres, Va = vascular bundles, o = outer, m = middle, i = inner and M = mean. The data are percentages.
The original article contains much more data, too much in the opinion of the present author to be cited here completely.

2. Thermal expansion

Keyword: Thermal expansion.
Subject: The coefficient of thermal expansion.
Source: Cox and Geymayer. Expedient reinforcement for concrete for use in S.E. Asia. 1969. Cited here is table 8.

Results: Three testmethods have been used. All bamboo specimens had the same diameter, about 19 mm.

Method 1: The mechanical strain was measured with pairs of diametrically opposed standard measuring disks for mechanical strain gauges of 200 mm apart parallel to the fibres. The specimens were coated with wax. Readings were taken at a temperature of 38 degree Celsius inside a constant-temperature room, and at -2 degree C. outside in the open, with about 3 hours exposure each.
Method 2: Strain gauges of 76 mm parallel to the fibres and 6 mm across the fibres; coating with wax. Specimens were placed in a temperature controlled oil-bath, at 10 and 38 degree C resp., each for several hours.
Method 3: Readings as in method 2. In addition, the specimens were sealed in steel tubes to keep them straight and to eliminate warping, which was suspected to have occurred during test methods 1 and 2. The results are as follows.

test method	n	MC %	coeff. of thermal expansion across	parallel
1	3	m=12.18	-	m=4.12
		s= 1.49	-	s=0.27
2	4	m=14.54	m=54.54	m=4.07
		s= 0.47	-	s=0.86
3	5	m=13.73	m=42.66	m=2.56
		s= 0.83	s= 3.80	s=1.17

in which n = number of specimens; MC = moisture content in %; m = mean; s = standard deviation; coefficient of thermal expansion is given in one millionth per degree, across the fibres and parallel to the fibres.

3. Moisture content and mass per volume

MOISTURE CONTENT 1

Keyword: Moisture content, position.
Subject: The moisture content for Bambusa vulgaris and Bambusa tulda in green condition, for bottom/middle/top.
Source: Talukdar, Y.A. and M.A. Sattar. Shrinkage and density studies on two bamboo species. Bano Biggyan Patrika, vol 9 nos 1/2, 1980, publ. 1982, pp 65-70.

Results: The authors have done tests on shrinkage (not reported here), mass per volume and moisture content on Bambusa vulgaris and Bambusa tulda from Bangladesh.
"Three mature and three immature culms of each species were collected from selected bamboo clumps of Chittagong and Chittagong Hill Tracts. The maturity of the bamboos was determined arbitrarily from their positions in the clump. Bamboos standing inside the clump were taken to be matured while those growing outside the clump were considered to be immature."
Samples for the determination of the mass per volume and the moisture content were taken in the form of 25 mm long rings from each of the height positions. Results are as follows.

Moisture content in %

position	maturity	Bamb.vulg.	Bamb.tulda
bottom	mature	52.8	77.2
	immature	85.7	90.2
middle	mature	48.7	75.9
	immature	86.3	75.2
top	mature	51.3	75.4
	immature	94.5	66.4
mean	mature	50.9	76.8
	immature	88.8	77.3

An analysis of variances did not show any correlation between moisture content on one hand and position, maturity or species on the other.
For data on mass per volume from this source, see p 15ff.

MOISTURE CONTENT 2

Keyword: Moisture content, position.
Subject: The moisture content of different parts of a bamboo culm, related to the height from bottom to top.
Source: Zhou Fangchun, Studies on physical and mechanical properties of bamboo wood. Journal of Nanjing Techn. Coll. of For. Prod., 1981 no 2 pp 1-32. In Chinese; the following is based on an English translation of the headings of the tables and the captions of the figures only. Cited here is table 15.

Results: The moisture content MC in % is related to the position along the length of the culm. For the purpose of a regression formula the length H is divided into ten parts, in which 0 means the bottom and 10 means the top. The regression formula reads:

$$MC = 94.5 - 12.7*H + 1.6*H^2 - 0.88*H\verb|^|3.$$

Table with data:

H	(1)	(2)	(3)
0	97.10	94.50	100
1	77.78	83.31	88.2
2	74.22	74.80	79.5
3	70.52	68.42	72.4
4	66.02	63.67	67.4
5	61.52	60.00	63.5
6	56.58	56.89	60.2
7	52.81	53.82	56.9
8	48.84	50.24	53.2
9	45.74	45.65	48.3
m	65.12	65.13	

in which H = the length of the culm, divided into 10 parts, 0 means the bottom and 10 means the top,
m = the mean value,
(1) = the measured moisture content MC in %,
(2) = the MC according to the regression formula,
(3) = idem, but related to 100.
Note: evidently this research is dealing with green bamboo.

MOISTURE CONTENT 3

Keyword: Age, moisture content.
Subject: The moisture content in Phyll.glauca cut at ages from 1 to 7 years.
Source: Lu Xiu-xin, Wang Ke-qing, Yi Xiang-chen, Liou Jin and He Ji-xun. A study on the physico-mechanical properties of culmwood of Phyll. glauca of Shandong. J. of bamboo res., Zhejiang For. Inst., Hangzhou, China, vol 4 no 2 (July 1985) pp 98-106.

Results: The authors have done tests on 54 culms of 4 different regions in the Shandong province; age classes from 1 to 7 years, and two culms per region and yearclass. Tests have been done on moisture content, mass per volume, tension, compression and bending, to determine the influences of the age at time of cutting and the region of origin. The number of tests was over 200 for each test and each region. The results dealing with moisture content are as follows.

region age	Sancha	Dajinkou	Dahuaya	Luchanya
1 year	95.66	88.58	95.53	121.22
2	87.39	86.85	91.96	105.25
3	79.02	84.97	80.26	99.11
4	77.46	84.73	79.45	76.18
5	77.21	81.54	70.67	74.02
6	73.21	80.96	67.87	70.40
7	66.93	71.77	66.52	

Moisture content in percent of dry weight.

MASS PER VOLUME 1

Keyword: Mass per volume.
Subject: The mass per volume in green, airdry and ovendry condition for 14 Taiwan bamboos.
Source: Chiang, F.C. Experiment on the physical properties of important bamboos in Taiwan. Bull. of the Taiwan For. Res. - Inst., no 241, July 1973, 12 pp.

Results: From specimens collected at the Liu-Kuei branch station of the Taiwan forestry institute, the weight has been determined in green, airdry and ovendry condition. The volume has been measured by immersion in water. The mass per volume of the bamboo substance has been determined (after grinding into dust) with a pycnometer. Results are as follows.

species	mass per volume in kg/m³			
	green	air-dry	oven-dry	sub-stance
Leleba oldhami Nakai	1070	540	430	1520
	20	80	50	320
Sinocalamus latiflorus McClure	1150	760	650	1530
	10	50	30	100
Phyll. edulis H. de Lehaie	1170	960	940	1670
	10	40	50	120
Leleba dolichoclada Odashima	1110	640	600	1370
	10	30	30	100
Leleba vulg. var. striata Nakai	1120	700	630	1550
	110	90	60	120
Phyll. lithophila Hayata	1150	870	840	1530
	60	70	50	90
Leleba pachinensis Nakai	1040	450	360	1480
	20	50	30	180
Leleba edulis Odashima	1130	950	870	1470
	30	40	50	90
Leleba dolichomerithalla Nakai	1160	800	750	1500
	20	40	30	160
Leleba vulgaris Nakai	1110	760	720	1470
	60	100	70	170
Bambusa stenostachya Hackel	1060	480	440	1550
	20	30	30	110
Phyll. makinoi Hayata	1210	830	750	1440
	20	30	20	90
Bamb. beecheyana Munro var.pub.Lin	1130	740	450	1410
	20	60	20	60
Bamb.dolich.Hayata Cv."Stripe"Lin	1090	440	400	1370
	20	460	40	140

Note: for each species the first line of data gives the mean value, and the second line gives the scatter; e.g. 1070 and 20 should be read as "1070 plus or minus 20".

The original article gives also data on the mass per volume, determined as the ovendry weight divided by the green volume, and the shrinkage.

MASS PER VOLUME 2

Keyword: Moisture content, position.
Subject: The moisture content for Bambusa vulgaris and Bambusa tulda in green condition, for bottom/middle/top.
Source: Talukdar, Y.A. and M.A. Sattar. Shrinkage and density studies on two bamboo species. Bano Biggyan Patrika, vol 9 nos 1/2, 1980, publ. 1982, pp 65-70.

Results: The authors have done tests on shrinkage (not reported here), mass per volume and moisture content on Bambusa vulgaris and Bambusa tulda from Bangladesh.
"Three mature and three immature culms of each species were collected from selected bamboo clumps of Chittagong and Chittagong Hill Tracts. The maturity of the bamboos was determined arbitrarily from their positions in the clump. Bamboos standing inside the clump were taken to be matured while those growing outside the clump were considered to be immature."
Samples for the determination of the mass per volume and the moisture content were taken in the form of 25 mm long rings from each of the height positions. Results are as follows.

Moisture content in %

position	maturity	Bamb.vulg.	Bamb.tulda
bottom	mature	52.8	77.2
	immature	85.7	90.2
middle	mature	48.7	75.9
	immature	86.3	75.2
top	mature	51.3	75.4
	immature	94.5	66.4
mean	mature	50.9	76.8
	immature	88.8	77.3

An analysis of variances did not show any correlation between moisture content on one hand and position, maturity or species on the other.
For data on moisture content from this source, see p 12.

MASS PER VOLUME 3

Keyword: Mass per volume, position.
Subject: The mass per volume, related to the position along the culm.
Source: Zhou Fangchun, Studies on physical and mechanical properties of bamboo wood. Journal of Nanjing Techn. Coll. of For. Prod., 1981 no 2 pp 1-32. In Chinese; the following is based on an English translation of the headings of the tables and the captions of the figures only. Cited here is table 16 and figure 7.

Results: The author gives data on the mass per volume rho in kg/m^3, related to the relative height H along the culm, divided into ten parts, 0 means the bottom and 10 means the top. The samples have been collected from Xiashu province, number of tests is 50. The regression formula is:

$$\text{rho} = 596 + 15*H + 0.4*H^2.$$

Data on rho:

part of the culm, H	0	2	4	6	8
measured rho	593	638	649	708	740
rho from formula	596	628	662	700	742

MASS PER VOLUME 4

Keyword: Mass per volume, position.
Subject: The mass per volume for soaked and airdry bamboo, related to the position across the wallthickness of the culmwall.
Source: Duff, C.H. Bamboo and its structural use. The Eng. Soc. of China, session 1940-41, paper no I.C.E. 1, Inst. of Civ. Eng., Shanghai 1941, pp 1-27. Cited here is figure 7.

Results: The author gives a diagram with the mass per volume in kg/m^3 plotted across the thickness of the culmwall. In this diagram, copied below, is the vertical axis the mass per volume in kg/m^3, upper curve soaked bamboo, lower curve dry bamboo. The horizontal axis represents the thickness of the culmwall, the left (0.00) is the outside, the right (72.00) is the inside. The numbers at the bottom are on an arbitrary scale.

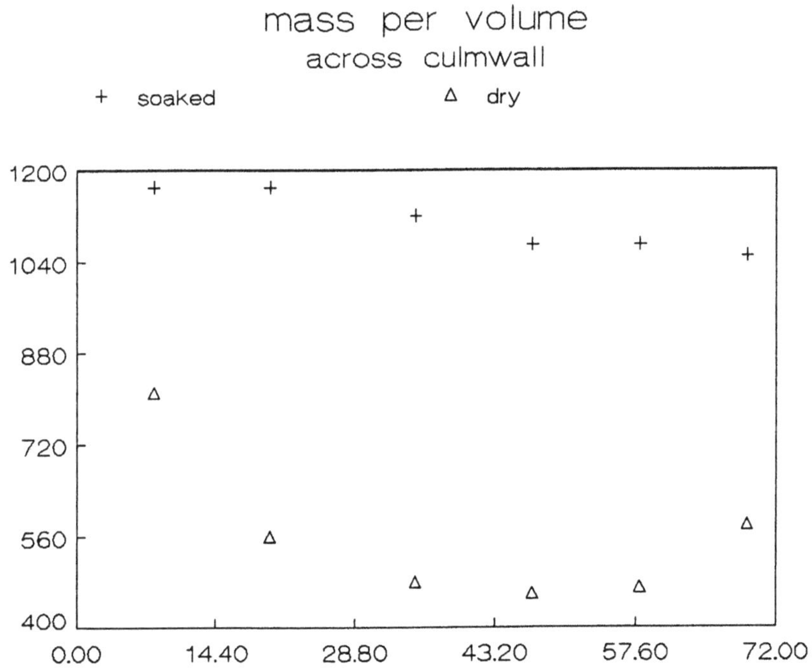

MASS PER VOLUME 5

Keyword: Age, mass per volume.
Subject: The mass per volume in Phyll.glauca cut at ages from 1 to 7 years.
Source: Lu Xiu-xin, Wang Ke-qing, Yi Xiang-chen, Liou Jin and He Ji-xun. A study on the physico-mechanical properties of culmwood of Phyll. glauca of Shandong. J. of bamboo res., Zhejiang For. Inst., Hangzhou, China, vol 4 no 2 (July 1985) pp 98-106.

Results: The authors have done tests on 54 culms of 4 different regions in the Shandong province; age classes from 1 to 7 years, and two culms per region and yearclass. Tests have been done on moisture content, mass per volume, tension, compression and bending, to determine the influences of the age at time of cutting and the region of origin. The number of tests was over 200 for each test and each region. The results dealing with mass per volume are as follows.

Test specimens were put into water for 2 weeks for maximum absorption. The volume has been determined by measuring the weight inside and outside the water. Finally specimens have been dried in an oven at 103 centidegrees. Results in kg/m^3:

region age	Sancha	Dajinkou	Dahuaya	Luchanya
1 year	583	556	554	547 kg/m^3
2	604	543	580	587
3	695	696	656	629
4	693	752	666	739
5	768	656	720	735
6	663	651	647	676
7	692	658	687	

The next regression formulae have calculated; a = age in years.

region	mass per volume in kg/m^3	st.dev.	best age
Sancha	484.9 + 93.4*a - 9.4*a^2	46	4.99
Dajinkou	446.7 + 106.1*a - 11.2*a^2	66.1	4.7
Dahuaya	475.2 + 77.2*a - 7.0*a^2	34.8	5.5
Lunchanya	504.7 + 119.1*a - 12.1*a^2	48.9	4.91

MASS PER VOLUME 6

Keyword: Age, mass per volume.

Subject: The mass per volume related to the age of the bamboo culms at the time of cutting.

Source: Zhou Fangchun, Studies on physical and mechanical properties of bamboo wood. Journal of Nanjing Techn. Coll. of For. Prod., 1981 no 2 pp 1-32. In Chinese; the following is based on an English translation of the headings of the tables and the captions of the figures only. Cited here is table 9 and figure 5.

Results: The author gives the following data on the mass per volume rho in kg/m^3:

<table>
<tr><td>age class A</td><td></td><td colspan="2">1</td><td colspan="2">2</td><td colspan="2">3</td><td colspan="2">4</td><td colspan="2">5</td></tr>
<tr><td>age in years</td><td>0</td><td>1</td><td>2</td><td>3</td><td>4</td><td>5</td><td>6</td><td>7</td><td>8</td><td>9</td><td>10</td></tr>
<tr><td>district:</td><td></td><td></td><td></td><td></td><td></td><td></td><td></td><td></td><td></td><td></td><td></td></tr>
<tr><td>Xiashu</td><td>243</td><td>425</td><td>558</td><td>608</td><td>626</td><td>615</td><td>630</td><td>624</td><td>657</td><td>610</td><td>606</td></tr>
<tr><td>Yixing</td><td></td><td colspan="2">574</td><td colspan="2">610</td><td colspan="2">635</td><td colspan="2">623</td><td colspan="2">619</td></tr>
<tr><td>Shimen</td><td></td><td colspan="2">543</td><td colspan="2">585</td><td colspan="2">637</td><td colspan="2">618</td><td colspan="2">563</td></tr>
<tr><td>Damaoshan</td><td></td><td colspan="2">506</td><td colspan="2">573</td><td colspan="2">613</td><td colspan="2">620</td><td colspan="2">608</td></tr>
<tr><td>mean</td><td></td><td colspan="2">529</td><td colspan="2">596</td><td colspan="2">627</td><td colspan="2">626</td><td colspan="2">600</td></tr>
</table>

The number of tests was 100 for each district, 400 in total. The author gives regression formulae for each of the four districts of origin, and for the mean. This last formula reads for the mass per volume rho in kg/m^3:

$$rho = 435 + 110*A - 15.5*A^2$$

in which A is the age class, see the heading of the table. The best age is 7.1 years. Evidently this formula and this best age are valid for the said regions only, but it shows how to determine the best age in another case.

MASS PER VOLUME 7

Keyword: Age, form of the culm, mass per volume and mechanical properties.
Subject: The dimensions of bamboo culms, the mass per volume and the mechanical properties related to the age.
Source: Sekhar, A.C. and R.K. Bhartari. A note on the strength of dry bamboo (Dendrocalamus strictus) from Madhya Pradesh. Indian Forester (1961, Oct.) pp 611-613.

Results: Material was collected from two localities in the said region. Pieces were about 3.30 m long, and in each age group 20 pieces were collected. Clear specimens only have been tested, conditioned to about 12% MC. Results:

age year	rho kg/m^3	Li mm	D mm	t mm	d mm
1	646	308	36.8	8.4	20.0
2	703	320	38.6	8.1	22.3
3	718	340	38.9	7.4	24.1
4	706	332	41.4	8.9	23.6
5	672	321	42.4	8.9	24.6
6	608	285	37.8	9.9	18.0

in which:
age is the age in years at the time of cutting,
rho is the mass per volume in kg/m^3,
Li is the average length of the internodes in mm,
D is the outside diameter of the culm in mm,
t is the wallthickness, and d is the inside diameter.

One should expect some relationship between the given data and the age, but it seems to be difficult to find any. Only the mass per volume appears to be the best at the age of three years.
Tests on static bending, impact bending and compression have been carried out also on these specimens.

ANATOMY AND MASS PER VOLUME 1

Keyword: Anatomy, mass per volume.
Subject: The influence of the structural elements on the mass per volume for different parts of bamboos.
Source: Ota, Motoi. Studies on the properties of bamboo stem (part 7). The influence of the percentage of structural elementson the specific gravity and compressive strength of bamboo splint. Bull. Kyushu Univ. For., vol 19 (1950), pp 25-47; tables 2, 5.

Results: For two botanical species, Ma-dake (Phyll. reticulata) and Moso-chiku (Phyll. edulis) Motoi Ota has determined the percentages of anatomical elements and their influence on the mass per volume. First the percentages of anatomical tissues:

Species and part	no of tests	outer layer	bundle sheath	fundamental tissue	inner layer
Phyll.ret					
outer	16	2.7	39.9	57.4	---
middle	8	---	25.9	74.1	---
inner	16	---	16.6	76	7.4
whole	8	1.3	28.4	66.6	3.7
Phyll.ed.					
outer	12	3.1	33.9	63.0	---
middle	6	---	22.8	77.2	---
inner	12	---	12.9	76.1	11.0
whole	7	1.3	22.2	71.4	5.0

With "bundle sheath" presumably the vessels and fibres are meant, and with "fundamental tissue" the parenchyma.

Next, the relationship between these anatomical elements and the mass per volume has been calculated:

species	tissue	correlation coefficient with mass per volume (rho)	regression equation
Phyll.ret	outer = O	0.121	rho = 896 + 6.6*(O - 2.2)
	bundle= V	0.655	826 + 6.0*(V -27.9)
	fund. = P	-0.720	826 - 7.4*(P -67.9)
	inner = I	-0.626	784 - 2.5*(I - 6.2)
Phyll.ed.	outer = O	0.781	1005 + 3.3*(O - 2.5)
	bundle= V	0.835	951 + 6.7*(V -23.0)
	fund. = P	-0.832	951 - 9.2*(P -71.2)
	inner = I	-0.726	911 - 1.4*(I - 8.8)

Keyword: Age, form of the culm, moisture content, mass per volume, fibre.

Subject: Length and diameter of fibres, and mass per volume for different species, related to age.

Source: Ku, Y.C. and C.H. Chiou. Tests on fiber morphology and chemical composition of important bamboos in Taiwan. Cooperative bulletin of Taiwan For. Res. Inst. no 20, publ. by Taiwan For. Res. Inst., Taipei, Febr. 1972, 8 pp.

Results: The authors have tested seven species, and determined data as follows.

spec.	age	H	DBH	MC	rho	fibre length			fibre width			l/w
		m	mm	%	kg/m³	max	min	mean	max	min	mean	
Phyll.	1	11.5	58	47.3	709	4.4	0.9	2.40	34	9	20.6	116
makinoi	2	11.2	53	42.3		5.4	1.1	2.86	34	7	17.6	162
Hay.	<3	10.8	52	38.8		4.3	1.1	2.47	60	9	18.1	137
sinocal.	1	18.0	117	69.6	459	5.1	1.0	2.82	26	4	14.6	193
latifl.	2	17.5	107	55.0		5.0	0.9	3.06	43	9	21.1	145
McClure	<3	15.2	91	40.9		6.0	1.0	3.09	37	7	18.1	171
Bambusa	1	18.6	95	66.6	601	3.7	1.6	3.12	26	7	14.2	220
stenost.	2	18.0	90	45.9		5.5	1.1	3.25	30	4	13.3	244
Hackel	<3	16.6	84	42.2		4.4	1.1	2.60	30	4	12.5	208
Leleba	1	11.0	59	56.6	729	4.4	1.1	2.81	39	4	14.2	198
dolich.	2	11.2	60	44.1		4.6	1.0	2.98	30	4	13.8	216
Odash.	<3	10.5	53	38.8		4.8	1.1	3.11	30	4	15.9	196
Phyll.	1	8.2	66	43.7	721	4.0	0.9	2.46	24	7	12.9	191
edulis	2	8.0	62	41.1		4.7	1.0	2.96	30	9	16.3	181
	<3	7.2	64	36.6		4.5	1.0	2.56	27	11	16.3	157
Leleba	1	9.0	51	61.3	671	3.7	0.8	2.43	41	4	19.8	123
oldhami	2	8.0	47	48.7		4.5	0.8	2.68	32	7	18.1	148
Nakai	<3	7.2	47	47.9		4.8	1.4	2.76	50	9	16.3	169
Bambusa	1	16.8	99	63.9	734	4.7	1.4	3.13	24	12	18.3	171
beech.M.	2	13.3	82	51.9		6.0	1.8	3.88	34	17	25.9	150
var.pub	<3	13.2	102	38.8		6.5	1.0	2.94	27	9	19.0	155

in which spec. = species, age is in years,
H = height of culm in m,
DBH = diameter at breast height in mm,
MC = moisture content of green bamboos in %,
rho = mass per volume in kg/m³,
fibre length is in mm, maximum/minimum/mean resp.,
fibre width is in μ = 0.001 mm, max./min./mean resp.,
l/w = ratio of fibre length and width.

This table gives valuable information on fibre dimensions; however also some remarkable data occur, e.g. the height H of the culms decreases with age, which obviously is not a physical fact. In the original report also data are given on the chemical composition of the bamboos.

ANATOMY AND MASS PER VOLUME 3

Keyword: Anatomy, mass per volume.
Subject: The number of vascular bundles related to the mass per volume.
Source: Zhou Fangchun, Studies on physical and mechanical properties of bamboo wood. Journal of Nanjing Techn. Coll. of For. Prod., 1981 no 2 pp 1-32. In Chinese, English translation of captures made by Eindhoven Univ. Cited here are tables 20 and 21.

Results: In the article by this author two tables can be found with a relationship between the number of vascular bundles and the mass per volume. First his table 21:

region of origin	vasc. n/mm²	rho kg/m³
Yixing	2.10	612
Xiashu	2.45	596
Shimen	2.09	589
Damaoshan	2.12	584

in which:
vasc.= number of vascular bundles per mm^2,
rho = mass per volume in kg/m^3.
The number of tests is 100 each, and 400 in total.
No correlation can be found.

Secondly his table 20:

site class of the forest	vasc. n/mm²	rho kg/m³
good	1.87	591
good/medium	1.96	597
medium	2.22	603
bad	2.40	602

from which can be calculated:

rho = 285 * vasc., or:

the mass per volume in kg/m^3 = 285 times the number of vascular bundles per mm^2. The variation coefficient is 10%.

4. Chemistry

CHEMISTRY 1

Keyword: Chemistry.
Subject: The chemical composition of some philippine bamboos.
Source: Tamolang, F.N., a.o. Properties and utilization of Philippine erect bamboos. Pp 189-200 in: Bamboo research in Asia. Proc. workshop Singapore May 1980. Editors Lessard, G., and Chouinard, A. Ottawa, ISBN: 0-88936-267-X.

Results: The chemical composition of some Philippine bamboos on a moisture free basis is as follows (all data in %).

species	Hol.	Pen.	Lig.	Alc.	Hot	NaOH	Ash	Sil
Gigant. levis	62.9	18.8	24.2	3.2	4.4	28.3	5.3	2.8
Schiz.lumampao	60.0	20.6	20.4	5.0	4.3	31.4	9.7	6.4
Gigant.aspera	61.3	19.6	25.5	5.4	3.8	22.3	4.1	2.4
B.vulgaris	66.5	21.1	26.9	4.1	5.1	27.9	2.4	1.5
B.blumeana	67.4	19.0	20.4	3.1	4.3	39.5	4.8	3.4
B.vulg.striata	63.6	21.5	25.9	3.7	3.9	24.7	3.0	1.3
10 Indian b.	-	15.1-21.5	22.0-32.2	0.2-3.2	3.4-6.9	15.0-21.8	1.7-3.2	0.44-2.1
10 Jap.etc.b.	61.9-70.4	17.5-22.7	19.8-26.6	0.9-10.8	5.3-11.8	22.2-29.8	0.8-3.8	0.1-1.7

in which:
Hol = holocellulose,
Pen = pentosans,
Lig = lignin,
Alc = solubility in alcohol-benzene,
Hot = idem in hot water,
NaOH= idem in 1% NaOH,
Ash = ash,
Sil = silica, all in %.
10 Indian b.= range of values for 10 Indian bamboo species,
10 Jap.etc.b. = range of values for 10 Japanese, Burmese and Indonesian bamboo species.

CHEMISTRY 2

Keyword: Fibre, chemistry.
Subject: The fibre dimensions and the chemical composition of six Taiwan bamboos.
Source: Chao, S.C., Y.C. Ku, S.J. Lin and T.T. Pan.
Measurements of fiber dimensions and analysis of chemical composition on Taiwan hardwoods. Bulletin of Taiwan For. Res. Inst., no 14, June 1971, 26 pp.

Results: The authors have done tests on fiber dimensions and chemical composition on numerous hardwoods, and on six bamboos as well. This report will deal only with the bamboos. The botanical species are: *Phyllostachys makinoi Hay; Dendrocalamus latiflorus Munro; Bambusa stenostachya Hackel; Bambusa dolichoclada Hay; Phyllostachys edulis Riviere, and Bambusa oldhami Munro.*
The test results are as follows.

species	Phyll. makinoi	Dendr. latifl.	Bambusa stenos.	Bambusa dolich.	Phyll. edulis	Bambusa oldhami
diameter in mm	52	91	84	53	64	47
fiber length						
in mm : max.	4.30	6.00	4.40	4.80	4.50	4.80
min.	1.10	1.00	1.10	1.10	1.00	1.00
mean	2.472	3.088	2.596	3.112	2.556	2.756
fiber width						
in mm : max.	0.060	0.037	0.030	0.030	0.027	0.050
min.	0.009	0.007	0.004	0.004	0.011	0.009
mean	0.0181	0.0181	0.0125	0.0159	0.0163	0.0163
length/width	137	171	208	196	159	169
wallthickness of cell in mm.						
max.	0.0153	0.0110	0.0088	0.0087	0.0131	0.0153
min.	0.0022	0.0022	0.0022	0.0022	0.0022	0.0022
mean	0.0063	0.0056	0.0052	0.0028	0.0044	0.0075
Chemical composition in %						
moisture	9.78	10.71	9.92	12.64	23.82	12.79
ash	1.51	2.82	1.29	1.19	1.12	2.17
extractives						
ether	0.52	0.85	0.82	0.57	0.46	0.44
hot water	4.25	5.77	5.93	6.06	8.60	3.84
1% NaOH	29.14	26.60	28.81	28.32	30.06	22.77
alcoholbenz.	2.78	7.37	4.20	8.37	4.01	5.61
pentosan	19.59	19.40	21.06	19.76	24.60	16.94
lignin	25.89	24.76	23.57	22.03	26.52	24.58
holocellulose	76.85	80.15	80.05	80.56	77.74	81.56
alphacellul.	48.63	48.14	48.27	46.65	45.69	48.77

The original article contains much information on the procedure of all the tests; unfortunately this is in Chinese only, without a summary in another language.

CHEMISTRY 3

Keyword: Age, chemistry, mass per volume, fibre.
Subject: The chemical and physical properties of Sinocalamus affinis related to the age at the time of cutting.
Source: Wu Meng Han Tuo, The cutting age of Sinocalamus affinis. Paper presented at the International Bamboo Workshop, Hangzhou, October 1985.

Results: The author has studied the chemical composition and the physical and mechanical properties for bamboo cut at different ages:

age	young	1 year	2 year	3 year	4 year	5 year
Chemical component in %:						
ash	1.49	3.46	1.85	1.46		2.51
SiO2	0.58	2.60	0.168	0.44		
1%NaOH extr.	34.82	27.81	24.93	22.91		26.89
total extr.(1)	13.77	9.22	6.60	6.04		11.19
lignin (2)	17.86	25.95	22.08	23.15		23.11
cellulose (3)	61.47	57.75	63.98	62.19		57.07
rho in kg/m^3	235	480	498	619	654	509
σ in N/mm^2		144	160	225	201	198
fibre:						
length in mm	1.95	1.81		1.95		1.95
width in mm	0.016	0.016		0.014		0.016
L/W	122	113		139		122

in which:
(1) = total extractives (alcohol-benzen: ethyl-alcohol and hot water).
(2) = lignin with 27% H_2SO_4 method.
(3) = cellulose with $NaClO_3$ method.
rho = mass per volume
σ = tensile stress.
In the opinion of the present author this research shows the importance of determining for each species the best age to cut the bamboo.

5. Elasticity

DYNAMIC VISCO-ELASTICITY

Keyword: Dynamic visco-elasticity, mass per volume.
Subject: Dynamic visco-elasticity related to mass per volume and outer or inner layer.
Source: Mamada, S. and Y. Kawamura. Ueber die Dynamische Viskoelastizitaet an der Bambusart (The dynamic visco-elasticity of bamboo). Mokuzai Gakkaishi (J. of the Japan Wood Res. Soc.) (1973) vol 19 no 11 pp 555-560.

Results: This article reports about the vibration reed method to determine the dynamic Young's modulus E and the loss factor (tan δ) for different species and outer or inner layer. Specimens have been tested with a MC of 11-12%. The size of the specimens was 50 mm long, 5 mm wide and 0.6 mm thick. The results are as follows.

	rho in kg/m³						E in N/mm²		tan δ	
species	outer			inner			outer	inner	outer	inner
	max	mean	min	max	mea	min				
Phyll.mitis	1110	1060	1000	780	740	700	15600	5800	0.0141	0.0223
Ph. bamb.	1110	1070	1020	650	580	520	20000	6500	0.0120	0.0158
Ph.nigra	1100	1070	1000	690	650	600	20700	7100	0.0128	0.0150
Pleiob.hind.	1220	1050	1000	550	510	470	22200	6500	0.0143	0.0145
Ph.takemuroi	1090	1040	950	550	490	460	19500	4800	0.0149	0.0170
Ph.bamb (1)	1120	1020	920	580	520	400	20000	5800	0.0156	0.0180
B.multiplex	990	930	870	580	520	480	19500	7100	0.0156	0.0161
B.vulgaris	950	860	800	500	450	500	11100	4500	0.0153	0.0208

(1) = form kashirodake.

The number of tests has been about 20 for each species. The dynamic E-value increases with increasing mass per volume, especially in the outer layer.

6. Bending

BENDING 1

Keyword: Bending, form of the culm.
Subject: Bending related to diameter and wallthickness.
Source: Espinosa, J.C. Bending and compressive strengths of the common philippine bamboo. Phil. J. of Sc., vol 41 no 2 (1930), pp 121-135, table 1.

Results: Espinosa has done 3-point bending tests on full culm, 110 tests in all, free span 1.5 m, airdry, speed of deformation 0.10 mm/s, botanical species Bambusa spinosa. He gives a table with the ultimate bending stresses as related to wallthickness and diameter. A summary of his table no. 1 reads:

d=	89.1	92.3	95.5	98.7	101.9	105.0	108.2	111.4
t=5	71.0	49.2						
6	59.5	56.9	62.4	50.9				
7		60.5	59.5	59.4	50.9			
8		39.1		55.3	50.1			
9		55.9		60.3		55.4		44.9

Content of this table: ultimate bending stresses in N/mm²,
d= outside diameter in mm,
t= wallthickness in mm.

An analysis shows that there is no relation between the said stresses on one hand and the diameter and/or the wallthickness on the other. The next figure shows clearly that the stresses are independent from diameter and wallthickness.

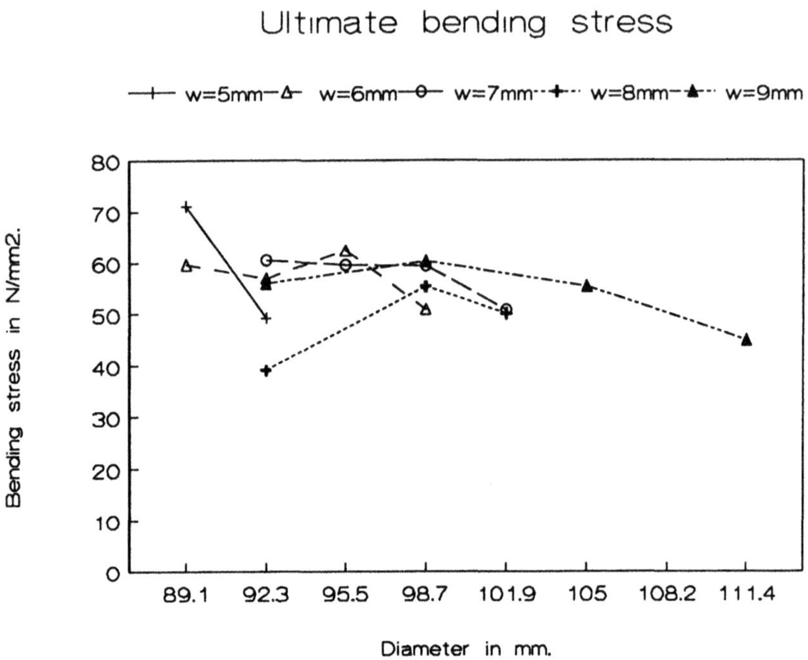
Ultimate bending stress
w=5mm
w=6mm
w=7mm
w=8mm
w=9mm
Bending stress in N/mm2.
80
70
60
50
40
30
20
10
0
89.1
92.3
95.5
98.7
101.9
105
108.2
111.4
Diameter in mm.

BENDING 2

Keyword: Bending, form of the culm.
Subject: Bending related to diameter and wallthickness.
Source: Espinosa, J.C. Bending and compressive strengths of the common philippine bamboo. Phil. J. of Sc., vol 41 no 2 (1930), pp 121-135, table 2.
Results: In a similar way Espinosa has done 145 bending tests on culms of a lesser diameter and a free span of 700 mm. A summary of his table no. 2 is:

d=	67	70	73.2	76.4	79.6	82.8	85.9
t=5	61.9	55.4	51.9	44.8	43	39.7	36.7
t=6	60.2			49	41.5	38.9	30.4

Content of this table: ultimate bending stresses in N/mm², d = outside diameter in mm, t = wallthickness in mm.

From this table it is clear that stresses are decreasing with increasing diameter;this is even more clear from the next figure.

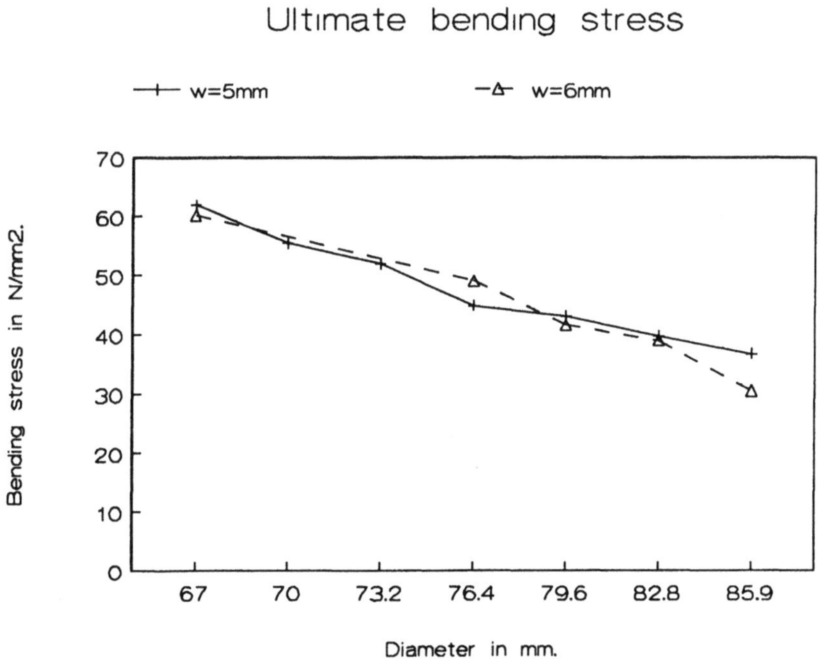

BENDING 3

Keyword: Bending, form of the culm.
Subject: Bending related to diameter and wallthickness.
Source: Espinosa, J.C. Bending and compressive strengths of the common philippine bamboo. Phil. J. of Sc., vol 41 no 2 (1930), pp 121-135, tables 1 and 2 together.

Results: From the bending tests on 1500 mm culms no relationship between stress and diameter could be seen; from the tests on 700 mm however such a relationship was clear. For that reason we plot them together in this figure:

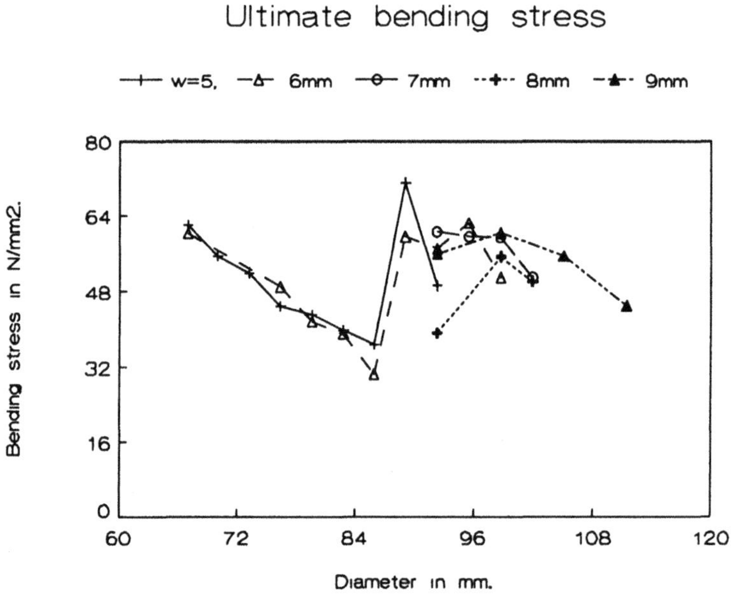

However, the left side of this figure represents the tests on a span of 700 mm, and the right side these on 1500 mm. Obviously this sample is not homogeneous, and the best fit is a horizontal straight line; the mean value is 51.3 N/mm^2, with a standard deviation of 8.7 N/mm^2.

BENDING 4

Keyword: Bending, age.

Subject: The ultimate bending stress for split bamboos, related to the age at the time of cutting.

Source: Zhou Fangchun, Studies on physical and mechanical properties of bamboo wood. Journal of Nanjing Techn. Coll. of For. Prod., 1981 no 2 pp 1-32. In Chinese; the following is based on an English translation of the headings of the tables and the captions of the figures only. Cited here are tables 12 and 13 and figure 6.

Results: The author gives the following data for the ultimate bending stress in N/mm^2, related to the age in years.

age class A	1	2	3	4	5	n
age in years	1-2	3-4	5-6	7-8	9-10	
district:						
Xiashu	106.5	126.3	120.8	139.4	119.2	45
Yixing	122.1	132.1	139.0	131.5	128.7	129
Shimen	103.2	120.3	123.4	128.8	122.5	146
Damaoshan	102.8	115.5	123.2	123.1	120.5	153
mean	108.7	123.6	126.6	130.7	122.7	473

in which n is the number of tests.

The author gives regression formulae for each of the four districts of origin, and for the mean. This last formula is:

$$\sigma = 89.6 + 22.6*A - 3.18*A^2.$$

The bending stress is = $(F*L/4)/(b*t^2/6)$ in which:

F = force, L = span = 240 mm, b = width of specimen = 10 mm, t = thickness of specimen = thickness of culmwall.

This position of the specimen with the outside and inside in a horizontal position is called "radial bending". The best age is 7.1 years for this case.

The author has also done tests on "tangential bending", with the outside and inside in a vertical position. In this case the stress is $(F*L/4)/(t*b^2/6)$. Results:

age class A	1	2	3	4	5	n
age in years	1-2	3-4	5-6	7-8	9-10	
district:						
Xiashu	87.2	146.3	140.3	142.0	123.7	112
Yixing	130.5	147.5	143.2	142.8	142.2	298
Shimen	116.7	127.2	133.0	139.4	130.8	467
Damaoshan	108.5	125.1	134.8	132.1	122.9	471
mean	110.7	136.5	137.8	139.0	129.9	1348

The regression formula for the mean is:

$$\sigma = 83.69 + 33.97*A - 4.98*A^2.$$

In this case the best age is 6.8 years.

BENDING 5

Keyword: Age, bending.
Subject: The ultimate bending stress in Phyll.glauca cut at ages from 1 to 7 years.
Source: Lu Xiu-xin, Wang Ke-qing, Yi Xiang-chen, Liou Jin and He Ji-xun. A study on the physico-mechanical properties of culmwood of Phyll. glauca of Shandong. J. of bamboo res., Zhejiang For. Inst., Hangzhou, China, vol 4 no 2 (July 1985) pp 98-106.

Results: The authors have done tests on 54 culms of 4 different regions in the Shandong province; age classes from 1 to 7 years, and two culms per region and yearclass. Tests have been done on moisture content, mass per volume, tension, compression and bending, to determine the influences of the age at time of cutting and the region of origin. The number of tests was over 200 for each test and each region. The results dealing with the ultimate bending stress have been determined with a three point bending test on split bamboos of 200 mm long, 5 mm wide and the wallthickness high. Results for the bending stress:

region age	Sancha	Dajinkou	Dahuaya	Luchanya
1 year	149	104	127	101 N/mm^2
2	160	128	144	136
3	178	152	153	151
4	183	163	168	197
5	170	153	158	190
6	166	147	159	170
7	178	149	163	

The next polynomials have been calculated (a = age in years):

region	ult.bending stress in N/mm^2	st.dev.	best age
Sancha	$\sigma = 136 + 16.9 \cdot a - 1.69 \cdot a^2$	10.4	4.97
Dajinkou	$74.5 + 35.0 \cdot a - 3.60 \cdot a^2$	8.14	4.85
Dahuaya	$110.5 + 20.0 \cdot a - 1.86 \cdot a^2$	6.32	5.38
Luchanya	$41.6 + 61.3 \cdot a - 6.50 \cdot a^2$	16.05	4.71

BENDING 6

Keyword: Age, mass per volume, bending.
Subject: The mass per volume and static bending related to the age.
Source: Sekhar, A.C. and R.K. Bhartari. A note on the strength of dry bamboo (Dendrocalamus strictus) from Madhya Pradesh. Indian Forester (1961, Oct.) pp 611-613.

Results: Material was collected from two localities in the said region. Pieces were about 3.30 m long, and in each age group 20 pieces were collected. Clear specimens only have been tested, conditioned to about 12% MC. Results:

age year	rho kg/m³	n	σ.el N/mm²	σ N/mm²	E N/mm²
1	646	14	57.0	96.2	16070
2	703	16	59.8	96.0	15300
3	718	16	56.0	91.7	15150
4	706	15	61.8	95.3	14470
5	672	17	55.3	93.6	14190
6	608	17	59.4	97.1	13730

in which:
age = the age in years at the time of cutting,
rho = the mass per volume in kg/m^3,
n = the number of tests,
σ.el= the stress at limit of elasticity in N/mm^2,
σ = the ultimate bending stress in N/mm^2,
E = the modulus of elasticity in N/mm^2.

Remarkably there is no relationship between the age and the given data; only E seems to decrease with age, which is difficult to understand.
Tests on these specimens have also been done referring to the size of the culm, impact bending and compression.

BENDING 7

Keyword: Bending, mass per volume.
Subject: The ultimate bending stress related to the mass per volume, for split bamboos.
Source: Zhou Fangchun, Studies on physical and mechanical properties of bamboo wood. Journal of Nanjing Techn. Coll. of For. Prod., 1981 no 2 pp 1-32. In Chinese; the following is based on an English translation of the headings of the tables and the captions of the figures only. Cited here is table 21.

Results: The author has done bending tests on split bamboos, with the outside and inside of the culm in a horizontal position, "radial bending", and with the outside/inside in a vertical position, "tangential bending". Results:

region of origin	rho	σ tang	n	σ rad.	n
Yixing	612	141.2	298	130.7	129
Xiashu	596	127.9	112	122.4	45
Shimen	589	129.4	469	119.6	146
Damaoshan	584	124.6	471	117.0	153

in which:
rho = mass per volume in kg/m^3,
σ = ultimate bending stress in N/mm^2,
tang = tangential, rad = radial, see before,
n = number of tests.
From these data one can derive:
tangential: $\sigma = 0.220*rho$ with a coeff. of variation V of 4%,
radial : $\sigma = 0.206*rho$, V = 3%.

BENDING 8

Keyword: Bending, age, position, mass per volume, seasoning.
Subject: Bending related to age, position in the culm and mass per volume.
Source: Limaye, V.D. Strength of bamboo (Dendr. strictus). Indian Forest Records, new series, timber mech., vol 1 no 1 (1952), 17 pp; reprinted from The Indian Forester, vol 78 no 11 (1952), pp 558-575.
Results: Limaye has collected a 200 culms, 50 of each age (0.5, 1, 2 and 2.5 years), and tested them on bending and compression in green and dry condition. The diameter was between 25 and 62 mm, and the length of the culms was between 5.4 and 7.8 m. According to diameter, wallthickness, length of culm and length of internode they were matched into pairs, and one of each pair was tested green and the other one dry.
From his table 1 follow the results on bending tests on green bamboos; three-point test, span 700 mm.

			node in centre					centre inter node				
age	p	n	MC	rho	σ.el	σ	E	MC	rho	σ.el	σ	E
0.5	B	11	111	491	30.3	55.8	9 000	113	522	34.5	59.3	11 800
	M	11	112	469	30.3	54.4	10 000	111	492	24.8	43.4	10 500
	T	11	106	473	28.9	56.5	12 300	101	475	26.9	44.1	13 200
	m	33	110	477	29.6	55.8	10 400	108	496	28.9	48.9	11 700
1	B	9	49	545	32.4	55.1	7 200	51	535	30.3	61.3	9 000
	M	9	61	493	28.2	53.7	9 100	71	530	24.8	49.6	9 600
	T	9	56	487	33.1	57.9	11 300	71	501	34.5	49.6	10 300
	m	27	55	508	31.0	55.8	9 000	64	522	28.9	53.7	9 500
2	B	10	51	593	44.1	74.4	10 000	48	628	47.5	78.5	11 100
	M	10	52	616	38.6	73.7	12 100	48	671	32.4	68.2	15 000
	T	10	49	616	37.9	69.6	12 500	45	688	45.5	68.9	15 600
	m	30	51	608	40.7	72.3	11 400	47	662	41.3	71.7	13 800
2.5	B	11	61	622	56.5	105.4	13 200	62	629	49.6	91.6	14 500
	M	11	55	676	58.6	98.5	13 600	57	665	48.9	84.7	14 800
	T	11	56	701	62.7	102.0	17 400	54	670	45.5	79.2	18 300
	m	33	57	666	59.3	102.0	14 700	58	655	48.2	85.4	15 800

in which
age = age in years
p = position in culm, B = bottom, M = middle, T = top, m=mean.
n = number of tests, the same quantity for each series.
MC = moisture content in percent
rho = mass per volume in kg/m^3
σ.el = stress at limit of elasticity in N/mm^2
σ = ultimate bending stress in N/mm^2
E = modulus of elasticity in N/mm^2
node in centre means: a node is in the centre of the span
centre inter node: the centre of the span is between two nodes

From the eight mean values m in this table we can easily calculate the next relationships, all related to the mass per volume in kg/m^3:

stress at elastic limit in N/mm^2	m = 0.0661	s = 0.0108
ultimate bending stress in N/mm^2	0.1147	0.0175
modulus of elasticity in N/mm^2	20.89	2.45

in which s means standard deviation.

Limaye gives in his table 3 also variances for the different sources of variation:

	DF	σ	E
B = between ages	3	3181 "	545 800 "
C = centre inter node vs node in centre	1	380	153 200
B*C	3	119	14 200
error (1)	-	59	31 000
(DF) =		(74)	(72)
D = between positions in the culm	2	182 "	293 000 "
B*D	6	3	21 600 '
C*D	2	99 "	4 300
B*C*D	6	11	9 900
error (2)	-	10	9 700
(DF) =		(147)	(122)

in which:
DF = degrees of freedom
" = significant at 0.1 percent level of probability
' = idem at 5 percent.

Limaye has done a lot of statistical calculations. A complete overview will be given now.

	σ	E
node in centre	71.7	11 500
centre inter node	64.8	12 820
standard error	1.9	420

in which σ = ultimate bending stress in N/mm^2,
E = modulus of elasticity in N/mm^2.

Next the influence of the position in the culm:

	bottom	middle	top	standard error
σ	73.0	65.5	66.1	1.0
E	10 750	11 850	13 850	290

These two influences together:

σ or E in N/mm²		σ			E	
bottom/middle/top	B	M	T	B	M	T
node in centre	73.0	70.3	71.7	9 850	11 230	13 370
centre inter node	73.0	61.3	60.6	11 580	12 470	14 330
standard error (each for B+M+T)		2.1			540	

Finally, the relationship of σ and E with the age in years:

age	σ	E
0.5	52.4	11 090
1	54.4	9 230
2	72.4	12 610
2.5	93.7	15 300
st.error	2.6	600

From these data one can calculate the next formulae:

$$\sigma = 38 + 20 \cdot age \quad \text{with } R = 0.96$$
$$E = 8520 + 2360 \cdot age \quad 0.84$$

and from the original data:

$$rho = 431 + 94 \cdot age \quad 0.97$$

in which σ = ultimate bending stress in N/mm^2,
E = elasticity modulus in N/mm^2,
rho = mass per volume in kg/m^3,
age = age in years,
R = square root of coeff. of determination.

As said before, Limaye has done bending test on matched pairs of bamboos, one of each pair was tested green and the other dry. The results of the dry tests are as follows.

			node in centre					centre inter node				
age	P	n	MC	rho	σ.el	σ	E	MC	rho	σ.el	σ	E
0.5	B	11	8.7	675	74.4	125.4	14060	10.0	695	59.3	108.2	16740
	M	11	8.6	687	71.7	124.0	15160	9.4	722	58.6	92.3	16540
	T	11	8.3	702	71.0	122.0	18330	9.5	692	54.4	81.3	16400
	m	33	8.5	688	72.3	124.0	15850	9.6	703	57.2	93.7	16540
			12.0			97.1	13710	12.0			80.6	15300
1	B	9	16.8	692	44.1	80.6	9300	15.6	690	48.9	85.4	12950
	M	9	14.6	701	51.0	84.7	10750	14.6	678	37.2	72.3	14680
	T	9	15.4	671	50.3	83.4	13160	13.7	634	44.8	71.0	13780
	m	27	15.6	688	48.2	82.7	11020	14.7	667	43.4	76.5	13780
			12.0			110.2	12950	12.0			91.6	16600
2	B	10	12.5	792	62.0	113.7	11160	13.5	785	59.9	125.4	14400
	M	10	9.7	832	76.5	120.6	14060	12.2	792	59.9	111.6	17360
	T	10	8.5	789	62.0	107.5	14610	11.3	806	66.1	111.6	19980
	m	30	10.2	805	67.5	113.7	13300	12.3	794	62.0	115.8	17290
			12.0			104.7	12950	12.0			115.8	17290
2.5	B	11	10.0	697	81.3	147.4	13640	10.3	748	82.0	143.3	18950
	M	11	9.3	753	95.1	153.0	16800	9.8	766	82.0	124.0	19360
	T	11	8.2	780	99.2	158.5	20330	9.6	796	86.1	124.0	21430
	m	33	9.2	743	91.6	153.0	16880	9.9	771	83.4	130.2	19910
			12.0			137.8	16330	12.0			118.5	19020

in which
age = age in years
P = position in culm, B = bottom, M = middle, T = top, m=mean
n = number of tests, the same quantity for each series.
MC = moisture content in percent
rho = mass per volume in kg/m^3
σ.el = stress at limit of elasticity in N/mm^2
σ = ultimate bending stress in N/mm^2
E = modulus of elasticity in N/mm^2
node in centre means: a node is in the centre of the span
centre inter node: the centre of the span is between two nodes

Limaye has calculated how σ and E improve with age for kilndry bamboo; for this purpose the MC has been adjusted to 12 percent.

age in years	0.5	1	2	2.5
σ in N/mm^2	88.9	100.6	110.2	128.2
E in N/mm^2	14540	14745	15160	17640

The improvement in strength due to seasoning is as follows.

age	MC	rho in kg/m³		σ in N/mm²		E in N/mm²	
	green	green	dry	green	dry(12%)	green	dry(12%)
0.5	110	487	696	52.4	88.9	11090	14540
1	60	515	678	54.4	100.6	9230	14740
2	49	635	800	72.3	110.2	12610	15160
2.5	58	661	757	93.7	128.2	15300	17640

BENDING 9

Keyword: Age, moisture content, position in culm, bending.
Subject: Static bending tests related to the age at the time of cutting, the moisture content and the position in the culm.
Source: Sekhar, A.C., B.S. Rawat and R.K. Bhartari.
Strength of bamboos: Bambusa Nutans. Indian Forester (1962, January) pp 67-73.

Results: The researchers have collected their material from a clump in the New Forest Estate in Dehra Dun. Forty new born half-year-old culms were selected, marked and numbered serially in December 1954. In July 1955 one-year-old culms numbered 1 to 8 were taken for tests, in July 1956 two-year-old culms numbered 9 to 16 were tested and so on till the five-year-old culms marked 33 to 40 were tested in 1959.

As soon as a culm was cut, two pieces of 50 mm length were taken, one with a node and the other without, from the bottom part to determine the intial M.C. After inspection each culm was divided into three parts: bottom, middle and top. Test procedure was as in Limaye's tests: three point bending test on a free span of 700 mm. The results are as follows (see next page).

		static bending, green						static bending, dry (12%)				
a	P	n	MC	rho dry	σ.el	σ	E	n	rho dry	σ.el	σ	E
1	B	8	122	516	49.0	82.8	9480	8	637	71.8	128.8	15380
	M	8	118	526	28.4	51.4	7570	8	630	54.1	80.4	11260
	T	8	117	520	26.7	49.9	11190	8	631	44.9	80.4	15180
	m	24	119	521	34.7	61.4	9410	24	633	56.9	96.5	13940
2	B	8	91	606	32.2	75.2	9420	8	734	54.6	108.0	11780
	M	8	87	622	22.1	45.3	6880	8	723	40.1	72.3	9430
	T	8	74	667	26.7	49.0	10200	8	728	37.9	68.6	12900
	m	24	84	632	27.0	56.5	8830	24	728	44.2	83.0	11370
3	B	8	97	593	44.8	77.5	9290	7	692	66.1	130.1	16200
	M	8	93	601	28.8	49.5	7600	7	674	40.8	70.3	10270
	T	8	84	626	31.4	49.7	11370	7	673	47.8	76.0	12760
	m	24	91	607	35.0	58.9	9430	21	680	51.6	92.1	13080
4	B	7	93	604	44.1	88.8	10690	7	737	55.5	104.6	12640
	M	7	91	619	33.9	63.2	10170	7	737	39.1	65.5	9220
	T	7	86	617	24.8	50.2	11660	7	729	34.1	58.1	11180
	m	21	90	613	34.3	67.4	10840	21	734	42.9	76.1	11010
5	B	8	82	667	58.0	97.2	12710	7	774	69.4	123.7	14290
	M	8	80	662	42.0	70.8	12130	8	775	48.3	85.1	14010
	T	7	85	637	33.0	63.1	14160	8	770	53.2	91.7	17870
	m	23	81	655	45.9	79.5	13010	23	773	56.6	100.2	15390

in which:
a = age in years,
P = position along the culm, B=bottom, M=middle, T=top, m=mean,
n = number of specimens, MC = moisture content in %,
rho = mass per volume in kg/m^3, dry = 12% MC,
σ.el = stress at limit of elasticity in N/mm^2,
σ = ultimate bending stress in N/mm^2,
E = modulus of elasticity in N/mm^2.

The present author has tried to find some correlations between these data. At first a correlation between the mass per volume and the ultimate stress was calculated, but the mass per volume was that widely scattered that no significant correlation could be found. The same remark holds for the E-modulus.
Next a comparison was made between the age and the position along the culm on one hand, and the ultimate stress or the E-modulus on the other hand. The result for the stresses is shown below; the plot for the E-modulus looks similar. The plot for the stresses gives some remarkable information: one should expect that stresses increase with age as well as from bottom to top, but in these results this is not the case.
Tests on compression, impact bending and shear have been done on specimens from the same culms.

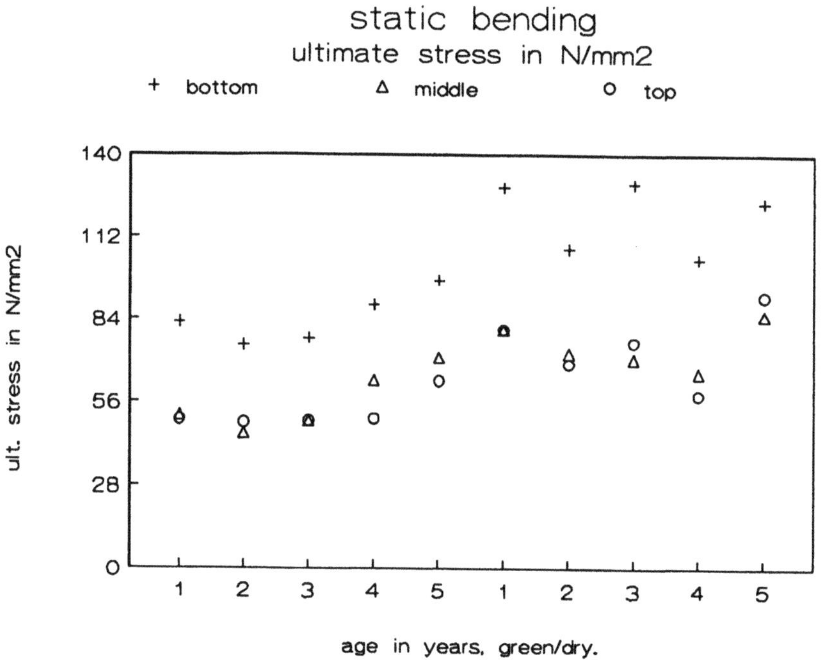

BENDING 10

Keyword: Bending, seasoning.
Subject: Bending related to species, and method of curing: wet or dry season, 30/60/90 days, in sun or in the shade.
Source: Teodoro, A.L. A preliminary study of the transverse strength of structural bamboo. Agric. Eng., vol 6 no 11 (1925, nov.) pp 266-267. Libr. E.T.H. Switzerland.

Results: Teodoro has tested 200 culms in bending, in a three point bending test with a free span of 3.00 m. Variables have been the botanical species: bambusa spinosa or vulgaris, both from the Philippines; air drying during the wet or the dry season, and lasting for 30, 60 or 90 days, and done in the sun or in the shade. The results are as follows, each line representing the mean values for about 15 to 18 specimens.

spec.	sea	j	ss	D	d	F	w	I	σ	E
B.sp.	dry	30	sh	81.7	63.4	3182	86.3	1.39	70	14900
			su	78.2	60.7	3002	97.6	1.17	75	14800
		60	sh	78.3	63.0	2440	101.0	1.07	67	12700
			su	85.5	69.4	2827	84.5	1.48	61	12700
		90	sh	84.2	63.7	2916	96.5	1.66	55	10200
			su	79.0	60.6	2447	98.5	1.25	58	11200
	wet	30	sh	77.3	62.6	2831	114.5	1.00	82	13900
			su	75.8	59.9	2616	105.5	0.99	75	14100
		60	sh	77.7	60.9	3331	118.0	1.11	87	14200
			su	79.0	64.5	2194	98.5	1.06	61	11800
		90	sh	80.0	62.0	3540	102.0	1.29	83	15200
			su	76.1	59.8	2870	96.5	1.02	80	16400
B.vu.	dry	30	sh	68.6	50.5	2182	108.7	0.77	73	14700
			su	71.0	55.0	2118	98.0	0.80	71	15200
		60	sh	75.1	59.0	1911	96.5	0.97	56	11500
			su	68.0	53.0	1743	114.0	0.66	67	13000
		90	sh	69.2	54.7	1831	81.0	0.69	69	18500
			su	64.2	48.0	1600	97.5	0.57	67	16100
	wet	30	sh	69.0	51.0	2236	124.0	0.78	74	13000
			su	67.5	51.4	1758	111.0	0.68	66	13200
		60	sh	75.0	59.0	2677	114.0	0.96	79	13800
			su	80.3	63.0	2550	109.0	1.27	61	10400
		90	sh	74.0	59.5	2499	106.0	0.86	81	15500
			su	70.6	55.0	2560	115.5	0.77	88	16200

in which: spec.= species, B.sp.= B. spinosa, B.vu.= B.vulgaris
sea. = season, dry or wet
j = number of days airdrying, 30, 60 or 90 resp.
ss = drying in the sun or shade; sh = shade, su = sun.
D = outside diameter in mm, d = inside diameter in mm.
F = maximum force in N; the ultimate force was higher or lower
w = deflection in mm occurring with F; the ultimate deflection (at failure) was more.
I = moment of inertia in 10^6 mm4, calculated from D and d.
σ = maximum bending stress, calculated from F.
E = Young's modulus, calculated from F and w.

An analysis of variances shows the next relationship:

$$\sigma = \beta 1x1 + \beta 2x2 + \beta 3x3 + \beta 4x4 + \beta 5x5 + \beta 6x6,$$ in which

$\beta 1x1$ represents the mean value for the stress σ,
x2 = +1 for B.vulgaris and -1 for B.spinosa,
x3 = +1 for the dry season and -1 for the wet season,
x4 and x5 represents the days of airdrying: for 30/60/90 days resp. x4 = -1/0/+1 and x5 = +1/-2/+1.
x6 = +1 for shade and -1 for sun. The result is:

$$\sigma = 71.08 - 0.08*x2 - 5.33*x3 - 0.31*x4 + 1.85*x5 + 1.92*x6.$$

Comments: x2 represents species; hardly of influence.
x3 represents the season; σ is 5.33 N/mm² better in the wet season and 5.33 worse in the dry season; difficult to understand.
x4 and x5: for 30 days +2.16 N/mm², for 60 days -3.7, and for 90 days +1.54 N/mm²; difficult to understand.
x6 makes σ 1.92 N/mm² higher in the shade, and 1.92 lower in the sun; due to cracking in the sun?
Similarly, for E is the formula:

$$E = 13900 + 375*x2 - 90*x3 + 345*x4 + 685*x5 + 125*x6.$$

Next a plot is given in which:
at the top: s = stress, sh = shade, su = sun, E = E-value,
at the left: stress in N/mm², or E in 100 N/mm²,
bottom: s = spinosa, v = vulgaris, d = dry or w = wet season, 3/6/9/= 30/60/90 days.

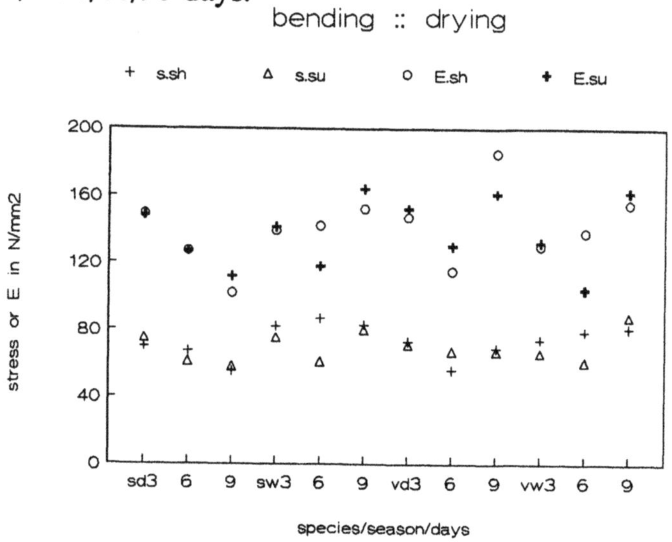

BENDING 11

Keyword: Bending, creep.
Subject: The increase of deformation with time in the case of bending for full culms.
Source: Atrops, J.L. Elastizitaet und Festigkeit von Bambusrohren (elasticity and strength of bamboo culms). Der Bauingenieur (1969) vol 44 no 6 pp 220-225. In German.

Results: The author has done four-point bending tests on full culms, free span 3.60 m, mean diameter 86 mm, mean wallthickness 9 mm ,number of tests 32, bending stress 16 N/mm^2.
Each 24 hours the load was removed for a moment, and the total deformation (the upper curve in the figure) and the remaining or plastic deformation (the lower curve) were measured. After 20 days, the culms were loaded until failure, at a stress of 80.2 N/mm^2, to be compared with a stress of 106.6 N/mm^2 without prior loading. The deformation is plotted in mm.

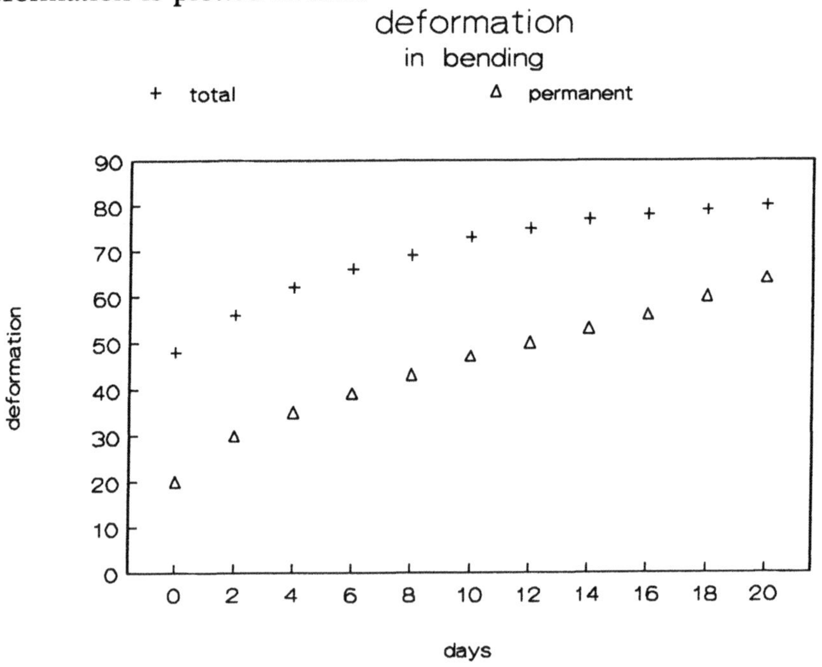

BENDING 12

Keyword: Bending, position in culm.

Subject: The ultimate bending stress for split bamboo, tested radial and tangential, with the position along the culm.

Source: Zhou Fangchun, Studies on physical and mechanical properties of bamboo wood. Journal of Nanjing Techn. Coll. of For. Prod., 1981 no 2 pp 1-32. In Chinese; the following is based on an English translation of the headings of the tables and the captions of the figures only. Cited here is table 18 and figure 7.

Results: The author has done bending tests on split bamboos, with the outside and inside of the culm in a horizontal position, "radial bending", and with the outside/inside in a vertical position, "tangential bending". He has divided the length of the culm into 10 parts, H = 0 means bottom, H = 10 means the top.
Results of the ultimate stress in N/mm²:

part of the culm, H	0	2	4	6	8
tang., 10 years old	100.1	115.6	125.4	141.3	163.6
tang., 6 years old	133.2	145.5	155.7	171.0	179.0
rad., 6 years old	114.5	134,7	139.6	140.8	151.0

The number of tests was 66, 155 and 43 resp. The specimens were collected from Shimen.

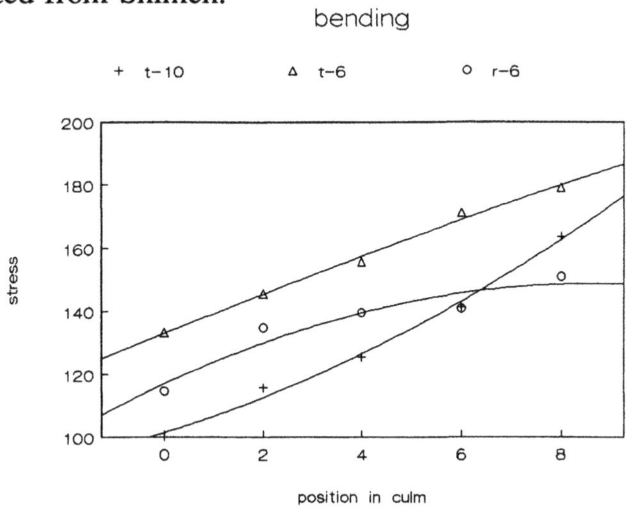

BENDING 13

Keyword: Bending.
Subject: Bending tests on split bamboos.
Source: Ueda, K. Mechanical properties of Moso bamboo (Phyll.pub). In: Res. Bull. Coll. Exp. For. Hokkaido Univ., vol 37 no 3 (1980), pp 817-836.

Results: The author has done tests on bending, compression and torsion on 51 specimens of Phyll.pub. The tests on bending deal with 33 specimens. The width of the split bamboos is between 8 and 11 mm, and the thickness or height is about 10 mm. The bending test is a three-point bending test with a free span of 170 mm. The bamboos have been tested in three positions: with the outer face towards the load, with the inner face towards the load, and radial. Results are as follows.

loading face		outer	inner	radial
rho in kg/m³,	min.	693	694	722
	mean	717	715	730
	max.	749	734	742
E in N/mm²,	min.	10160	9770	11570
	mean	10520	10120	12070
	max.	11100	10600	12650
σ.el in N/mm²,	min.	75.95	38.78	57.23
	mean	80.20	41.81	60.09
	max.	87.43	47.33	62.12
w at σ.el in mm,	min.	3.3	1.8	2.6
	mean	3.5	1.9	3.1
	max.	3.9	2.2	3.6
σ in N/mm²	min.	105.83	114.33	119.95
	mean	108.25	121.73	124.92
	max.	113.66	129.88	128.32
w at σ in mm,	min.	5.1	20.0	11.7
	mean	5.6	22.4	13.1
	max.	6.3	26.3	14.0

in which w = deflection,
σ.el = stress at limit of elasticity,
σ = ultimate stress.

It is interesting to see the difference in behaviour: loaded on the outer face, the bamboo behaves like a brittle material, because the soft innerside is at the tensile side. Loaded on the inner side, the hard and tough outer side is at the tension-side and consequently the behaviour is much better.
The original publication deals extensively with the differences between the E-modulus in the two tangential positions and the radial position (not quoted here).

BENDING, TESTMETHOD 1

Keyword: Bending, testmethod.
Subject: The ultimate bending stress for split bamboos, and for full culms.
Source: Atrops, J.L. Elastizitaet und Festigkeit von Bambusrohren (elasticity and strength of bamboo culms). Der Bauingenieur (1969) vol 44 no 6 pp 220-225. In German.

Results: The author has done four-point bending tests on split bamboos with a free span of 300 mm, and for two cases: the exterior of the bamboo on top, or down. The test-speed was 0.83 $N/mm^2/s$. Results:

	exterior on top		exterior down	
	w	σ	w	σ
min.	L/20.2	88.5	L/9.6	68.9
mean	L/13.8	142.5	L/7.7	113.4
max.	L/10.6	216.0	L/5.1	183.0

in which:
w = deflection in mid-span, just before failure,
L = free span = 300 mm,
σ = ultimate bending stress in N/mm^2.

The Young's modulus was:
exterior on top: 13600 to 19800 N/mm^2,
exterior down : 16900 to 22700 N/mm^2.

The test scheme for bending tests on full culms contains fout loads F and a free span of 3600 mm:

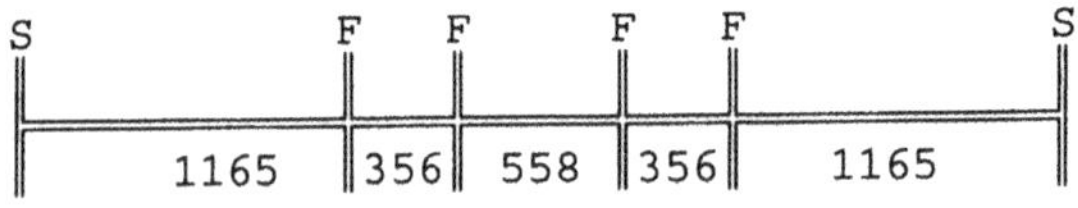

in which:
S means the position of a support, and F that of a force or load,

the numbers 1165 etc. mean the distances in mm between support and load, or between one load and the next; their total is 3600 mm, the free span.
Number of tests 40, diameter from 70 to 100 mm, wallthickness 6 to 12 mm. Results:

	w	σ
min.	L/25.9	76.1
mean	L/20.1	106.6
max.	L/16	143.3

in which w, L and σ as before (however, here is L = 3600 mm).

BENDING, TESTMETHOD 2

Keyword: Bending, testmethod.
Subject: Comparison of 3- and 4-point bending tests.
Source: Meyer, H.F. and B. Ekelund.
Tests on the mechanical properties of bamboo.
The Engineering Soc. of China. Session 1922-1923, vol 22 paper 7, pp 141-169. Present in library E.T.H. Switzerland.

Results: Certainly these are the oldest tests ever performed and published. Unfortunately the botanical species has not been mentioned, neither the diameter or the wallthickness. Bending tests have been done in three groups of 50 culms each: two groups with a 3-point bending test, on a span of 1.80 or 2.10 m resp. The ultimate bending stress was 90 N/mm^2. The third group was a 4-point bending test on a span of 2.10 m, with an ultimate stress of 96 N/mm^2. From figure 3 in the original article (not given here) it appears that the distribution is rather wide, and skew as well. According to the original author this is due to the variation in the mass per volume: 500 kg/m^3 for old dry bamboo, 1020 for green and 700 for seasoned bamboo. It looks like if all these bamboos have been used in these bending tests. The E-modulus is reported as 11 400 N/mm^2.

BENDING, TESTMETHOD 3

Keyword: Bending, testmethod.
Subject: Simple and effective method to test bamboo in bending.
Source: Teodoro, A.L. A preliminary study of the transverse strength of structural bamboo. Agric. Eng., vol 6 no 11 (1925, nov.) pp 266-267. Libr. E.T.H. Switzerland.

Results: Teodoro has developed a very clever method to test bamboo in bending. For that reason his text is reproduced here.
"The platform scale and levers were arranged as shown in the diagram. A V-shaped block "a" was placed right on the center of the platform scale. Another block "b" was placed at a distance of 2.7 metres from "a". On these blocks rested an I-beam of iron. On the beam 30 cm from the block "b" was placed another block "c". At a distance of 3 m from this last block, perpendicular to the beam was the jack screw. Still another block "d" similar to the block "c" was placed on the jack screw. On these last two blocks rested channeled rectangular blocks "rb", the lower surfaces of which were covered with iron. This gave sufficient strength to the blocks. All the edges of the blocks were covered with iron of about 0.63 cm thick. The edges of the first two blocks were made parallel also those of the other two blocks. The two pairs were perpendicular to each other.

The sections of the specimen rested on the channeled blocks. The height of the block on the jack screw was in its lowest position. An iron strap "B" as shown in the figure was used to supply the mechanical load. Another channeled block "Z" was placed on the section between it and the belt. The purpose of this was to distribute the load over the upper face of the section.....
The readings of the scale beam and measurements of deflections were taken at each change of load, the change of load being made by raising the jack screw. The increase of load was uniform until the elastic limit was reached. For convenience of readings, a uniform increment of load was used..... At each change of load, sufficient time was allowed to elapse, so that the motion had ceased. The jack screw was turned slowly until the desired load was

registered by the balance. In this way the elastic limit may be easily determined. The loading was then continued until the point of rupture was reached. No shocks or jerks were made. ... The average time required for testing a section was about fifteen minutes."

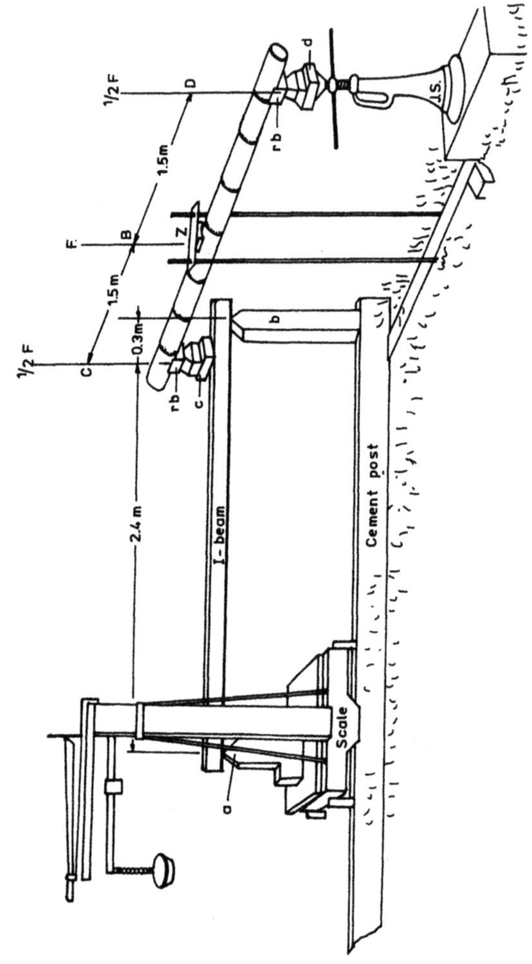

IMPACT BENDING 1

Keyword: Age, mass per volume, impact bending.
Subject: The mass per volume and impact bending related to the age.
Source: Sekhar, A.C. and R.K. Bhartari. A note on the strength of dry bamboo (Dendrocalamus strictus) from Madhya Pradesh. Indian Forester (1961, Oct.) pp 611-613.

Results: Material was collected from two localities in the said region. Pieces were about 3.30 m long, and in each age group 20 pieces were collected. Clear specimens only have been tested, conditioned to about 12% MC. Results:

age year	rho kg/m^3	n	σ.el N/mm^2	E N/mm^2
1	646	2	169.5	23560
2	703	8	170.2	29410
3	718	9	169.5	28730
4	706	13	153.4	26790
5	672	15	147.5	20290
6	608	6	136.1	18160

in which:
age = the age in years at the time of cutting,
rho = the mass per volume in kg/m^3,
n = the number of tests,
σ.el= the stress at the limit of elasticity in N/mm^2,
E = the modulus of elasticity in N/mm^2,
the last two according to ASTM for impact bending.

It is difficult to see a clear relation between age and the given data; the only conclusion could be that bamboo at the age of 3 years is at its best.
Tests have been carried out on the same specimens, dealing with the size of the culm, static bending and compression.

IMPACT BENDING 2

Keyword: Age, moisture content, impact bending.
Subject: Tests on impact bending for culms of different age and in green or dry condition.
Source: Sekhar, A.C., B.S. Rawat and R.K. Bhartari.
Strength of bamboos: Bambusa Nutans.
Indian Forester (1962, Jan) pp 67-73.

Results: Impact bending tests have been carried out on the species Bambusa Nutans, in green and dry condition, on the middle part of culms of different ages. Tests have been done with a swinging pendulum impact machine, on a free span of 700 mm.

Results are as follows.

	green				dry		
a	n	MC	σ.el	E	n	σ.el	E
1	8	114	47.7	5660	7	66.4	9030
2	8	82	53.8	6610	8	64.9	6570
3	8	87	54.7	6670	7	69.3	9515
4	7	96	72.8	11000	7	55.3	7145
5	8	82	103.9	20070	8	102.2	18570

in which a = age in years,
n = number of tests,
MC = moisture content in %,
σ.el = stress at limit of elasticity in N/mm^2,
E = E-modulus in N/mm^2.

In their publication from 1956 Sekhar and Rawat define the stress and E-modulus for impact bending as follows:

$$\sigma.el = 7130.13 * D * H / (D^4 - d^4) * w$$

$$E = 130.67 * \sigma.el / D * w$$

in which σ.el and E in psi,
D = outside diameter, d = inside diameter,
H = height of drop at elastic limit,
w = deflection at elastic limit, all in inches.

In ISO-units these formulae will become as follows:

$$\sigma.el = 49.13 * D * H / (D^4 - d^4) * w, \text{ and}$$

$$E = 84300 * \sigma.el / D * w, \text{ in which}$$

$\sigma.el$ = stress at limit of elasticity (according to ASTM) in N/mm^2
E = modulus of elasticity, idem,
D = outside diameter in mm, d = inside diameter,
H = height of drop at elastic limit in mm,
w = deflection at elastic limit in mm.

Tests on compression, static bending and shear have been done on specimens from the same culms.

7. Compression

COMPRESSION 1

Keyword: Compression, testmethod.
Subject: Compression related to the height of the specimen.
Source: Espinosa, J.C. Bending and compressive strengths of the common philippine bamboo. Phil. J. of Sc., vol 41 no 2 (1930) pp 121-135, tables 4, 5 and 7.

Results: Espinosa has carried out compression tests on airdry bamboos, 54 tests, height of the specimen 1200 mm, speed of deformation 0.02 mm/s, Bambusa spinosa. One end of the specimen was U-shaped "to take rounded bearing surface...as generally used in construction." In the test a piece of wood fits exactly the U-end. The other end was cut at right angles to the axis.

A summary of his table no. 4 is:

d=	102	105	111.4 mm
t = 9 mm	30.3	32	
10		28.8	26.4
11			27

Content of this table: ultimate compressive stresses in N/mm², d = outside diameter in mm, t = wallthickness in mm.

Next, he has done also tests on specimens of 350 mm height: (summary of his table no. 5)

d =	47.8	54.1	57.3	60.5	63.7	66.9	70.0
t = 4			82.8	65.8			
5	60.5	65.7	65.0	65.3	57.8	51.8	53.6

Explanation of this table: as before.

Finally, his table no. 7 gives the results of compression tests on split bamboos, height 25 mm, cross-section about 9 by 15 mm, moisture content 13 percent, n = 100. The mean ultimate compressive stress in this case is 53.5 N/mm².

COMPRESSION 2

Keyword: Compression.
Subject: Ultimate compression stress.
Source: McClure, F.A.
Some preliminary tests on the longitudinal crushing strength of Hua-Mei-Chu, a variety of Bambusa Tuldoides.
Lingnan Sc. J., vol 17 no 1 (April 8th, 1938), Lingnan Univ., Canton. The original is not available to the present author; only an abstract, made by Building Research Establishment, U.K., in 1942, B.R.S. Library Archive no. 161.

Results: McClure has done compression tests on bamboo from the market. The bamboo has been submerged for one or two months, and the pieces to be tested had been out of water for some days; "parts of them being alive and parts dead". Three culms were cut into pieces of 300 mm each, 7 pieces from the first culm, 3 from the second and 11 from the third, 21 specimens in all, 20 with a node and one an internode. Results:

	mean value	standard dev.	
diameter,max.	47.9	8.6	mm
diameter,min.	46.9	8.8	mm
wall thickness,max.	8.6	4.1	mm
idem,min.	8.0	3.6	mm
ultimate stress	35	8	N/mm^2

COMPRESSION 3

Keyword: Form of the culm, compression, position.
Subject: Mathematical formulae describing the diameter as function of the height, and the stress in the culm due to own weight.
Source: Watanabe, M. and S. Oohata. Studies on bamboo culm form, on Phyll. bamb. In: J. Jap. For. Soc. vol 62 no 1 (Jan. 1980) pp 9-16.

Results: The authors have done series of measurements on bamboo culms. Most of their results deal with e.g. the quantity of leaves; in the opinion of the present author these data are rather botanical and as such outside the scope of this book. Some data however are very interesting.

Figure 3 in the original paper gives the compression stress in the culm in the middle portion of each internode as function of the height. This stress is a constant: the value is about 0.09 N/mm^2. From the fact that this stress is a constant along the culm, we can understand how good the design of the culm is.

Next, figures 9 and 10 give relations between the diameter of the culm and the height of the culm. The external diameter at breast height D (in cm) and the culm height (in m) are related as follows:

$$H = 3.1832 * D \wedge 0.7563;$$

according to the present author the next formula does as well:

$H = 2.3 + 0.16 * D$, in which H in m and D in mm.

Figure 10 gives the same ratio as: $H = 3.527 * D \wedge 0.9478$,
instead of which the actual author proposes: $H = 0.33 * D$.
Finally the diameter at 0.1 of culm height is given as :

$H = 3.8074 * D$, in which H in m and D in cm,

or in the opinion of the actual author:

$H = 0.38 * D$, in which H in m and D in mm.

COMPRESSION 4

Keyword: Anatomy, compression.
Subject: The influence of the structural elements on the ultimate compression stress for different parts of bamboo.
Source: Ota, Motoi.
Studies on the properties of bamboo stem (part 7). The influence of the percentage of structural elements on the specific gravity and compressive strength of bamboo splint. Bull. Kyushu Univ. For., vol 19 (1950), pp 25-47; table 6.

Results: For two botanical species, Ma-dake (Phyll.reticulata) and Moso-chiku (Phyll. edulis), Motoi Ota has determined the relationship between the percentage of anatomical elements and the ultimate compression stress:

species	tissue	correlation coefficient with σ'	regression equation
Phyll.ret	outer = O	0.458	σ' = 104.8 + 6.9*(O - 2.2)
	bundle= V	0.865	88.1 + 1.8*(V -27.5)
	fund. = P	-0.728	88.1 - 1.9*(P -68.5)
	inner = I	-0.724	74.6 - 3.6*(I - 5.8)
Phyll.ed.	outer = O	0.940	96.3 +13.0*(O - 2.5)
	bundle= V	0.899	84.6 + 1.7*(V -23.0)
	fund. = P	-0.902	84.6 - 2.3*(P -71.2)
	inner = I	-0.196	74.1 - 1.7*(I - 8.8)

in which: outer = the percentage of outer layer,
bundle= idem of vessels and fibres,
fund. = idem of the parenchyma,
inner = idem of the inner layer.
σ' = the ultimate compression stress in N/mm^2.

COMPRESSION 5

Keyword: Anatomy, mass per volume, compression.
Subject: The influence of the outer and inner layer of the specimen on the mass per volume and the ultimate compression stress in split bamboo.
Source: Ota, Motoi.
Studies on the properties of bamboo stem (part 5). Study on the effect of the absence of the outer-layer or the inner-layer of bamboo splint on the compressive strength parallel to the grain and the specific gravity (air dry).
J. Jap. For. Soc. (1951), vol 33, pp 244-246.

Results: Motoi Ota has carried out tests on the compression strength of split bamboo specimens of the botanical species Phyll. reticulata and Ph. edulis. The main variable is if the thickness of the specimen is the full thickness of the wall of the bamboo culm, or the outer part only or the inner part only. The width of the specimen is always equal to the full thickness of the culmwall, and the height is twice that size. The results are as follows.

species		full thickness			inner part only			outer part only		
and no.		MC	rho	σ	MC	rho	σ	MC	rho	σ
Ph.re.	1	13.8	891	77	13.6	893	71	14.4	909	82
	2	12.8	877	74	12.5	882	73	12.4	895	82
	3	14.2	898	76	14.1	892	74	14.0	878	77
	4	13.5	970	86	14.2	967	83	13.5	940	86
	5	13.6	972	90	13.8	969	87	13.2	963	99
	m	13.6	922	81	13.6	921	78	13.5	917	85
	s	0.5	46	7	0.7	43	7	0.8	34	8
Ph.ed.	1	14.1	788	64	14.0	787	61	14.1	816	64
	2	16.3	735	52	16.9	756	52	17.0	727	52
	3	13.9	970	85	13.6	964	82	13.2	954	84
	4	14.6	832	64	14.2	811	54	14.7	824	60
	5	14.5	823	77	13.4	808	71	13.9	796	76
	m	14.7	830	68	14.4	825	64	14.6	823	67
	s	0.9	87	13	1.4	81	13	1.4	82	13

in which MC = moisture content in %
rho= mass per volume in kg/m^3
σ = ultimate compression stress in N/mm^2
m = mean value
s = standard deviation

In the opinion of the present author the differences are less than one should expect: the outer layer should be heavier and stronger, and the inner layer should have less mass per volume and less strength.

COMPRESSION 6

Keyword: Moisture content, compression.
Subject: The ultimate compression stress of Phyll.pub. related to moisture content.
Source: Li Yunlien and Li Yezhen. Physico-mechanical properties of culmwood of Phyll.pub. produced in Guizhou. Bamboo research, Nanjing Techn. Coll. of For. Prod. (1983), no 1, pp 52-74.

Results: In October 1980 90 culms have been collected from 21 fields. All culms were normal, with a diameter of 100 mm or more and a length of 15 m. Tests were done on specimens from the lowest 6 m. Compression test specimens were 20 mm to 20 mm to t mm (t = wallthickness), selected from 7 years old culms. Results:

MC in %	2.1	5.4	7.5	15.9	22.5	27.9	37.3	45.7	65.4
σ' N/mm²	110.4	96.9	83.9	64.5	60.3	57.2	55.8	56.0	54.3
n	11	10	10	10	12	12	12	12	11
s	7.04	7.45	6.69	4.90	4.31	3.11	3.24	2.75	4.12
st.err.	2.12	2.36	2.12	1.55	1.24	0.90	0.94	0.80	1.24
V in %	6.37	7.69	7.97	7.59	7.15	5.43	5.81	4.92	7.58
p	1.92	2.43	2.52	2.40	2.06	1.06	1.68	1.42	2.28

in which:
MC = moisture content in %,
σ' = ultimate compression stress in N/mm²,
n = number of tests,
s = standard deviation,
st.err. = standard error Sx,
V = variation coefficient,
p = accuracy index.

The least square method gives the next regression line:

$$\sigma' = 56.56 + 130/MC$$ with correlation factor r = 0.93.

With this formula the fibre saturation point can be calculated to be 30%.

COMPRESSION 7

Keyword: Compression, moisture content.
Subject: The ultimate compression stress related to the moisture content.
Source: Zhou Fangchun, Studies on physical and mechanical properties of bamboo wood. Journal of Nanjing Techn. Coll. of For. Prod., 1981 no 2 pp 1-32. In Chinese; the following is based on an English translation of the headings of the tables and the captions of the figures only. Cited here is table 8 and figure 4.

Results:

moisture content MC in %	3.80	7.98	14.11	24.11	77.98
no. of tests, n	10	10	10	10	10
ultimate compr. stress:					
mean, m, in N/mm²	77.7	67.1	58.15	45.9	45.1
st.dev. s in N/mm²	3.46	3.58	2.40	2.08	1.51
st.dev. of mean in N/mm²	1.096	1.132	0.76	0.66	0.48
variation coeff. V in %	4.46	5.33	4.13	4.54	3.35
relative error in m, P, %	1.41	1.69	1.31	1.44	1.06

The standard deviation of the mean = $s/\sqrt{n}$. The relative error = $(s/\sqrt{n}/m)*100\%$. Next follows a plot of these data.

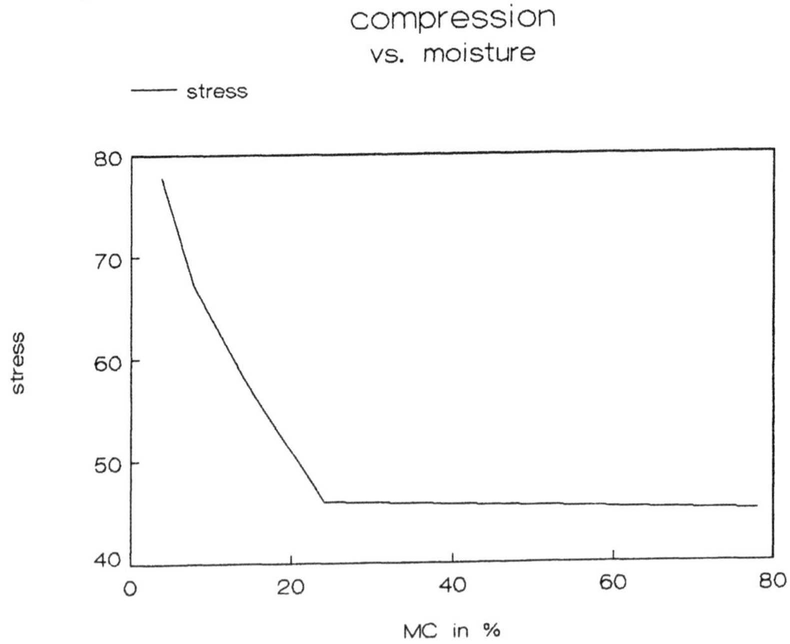

COMPRESSION 8

Keyword: Mass per volume, compresssion.
Subject: The influence of the mass per volume on the ultimate compression stress for different parts of bamboos.
Source: Ota, Motoi.
Studies on the properties of bamboo stem (part 7). The influence of the percentage of structural elements on the specific gravity and compressive strength of bamboo splint.
Bull. Kyushu Univ. For., vol 19 (1950), pp 25-47; tables 1 and 7 and figure 5.

Results: Botanical species Ma-dake (Phyll. reticulata) and Moso-chiku (Phyll. edulis) have been used in these tests. Specimens have been taken from the outer part, the middle part and the inner part of the wallthickness, more or less in the form of a cube, very small indeed. Compression tests have been done with a speed of 0.04 m/s. Results:

Species and part	no.of tests	MC in percent	mass per volume in kg/m³			ult.compr.stress in Nmm²		
			min.	mean	max.	min.	mean	max.
Phyll.re.								
outer	24	10.0-13.5	837	918	1041	92.9	112.2	130.8
middle	8	10.0-13.0	718	775	813	62.9	79.7	95.1
inner	24	10.5-13.6	629	748	833	52.3	67.2	81.6
whole	32	10.4-12.9	756	866	944	78.2	83.1	96.0
Phyll.ed.								
outer	21	11.5-14.0	957	1030	1076	95.6	105.1	111.3
middle	6	12.5-13.6	919	928	948	67.3	76.0	81.2
inner	21	12.0-14.7	813	883	938	57.9	68.4	76.0
whole	28	11.9-14.7	923	965	1012	69.9	81.6	89.6

Motoi Ota gives (in his figure 5 and table 7) correlations between the mass per volume and the ultimate compression stress for outer part, inner part and whole. It might be more interesting to calculate a correlation between the mean mass per volume and the mean compression stress for each species:

Phyll.ret.: $\sigma' = -94.5 + 0.218*rho$ with $R = 0.90$
Phyll.ed. : $\sigma' = -153.7 + 0.249*rho$ 0.98

in which:
σ' = ultimate compression stress in N/mm^2,
rho = mass per volume in kg/m^3.
R = square root of coeff. of determination.
A plot of these formulae is:

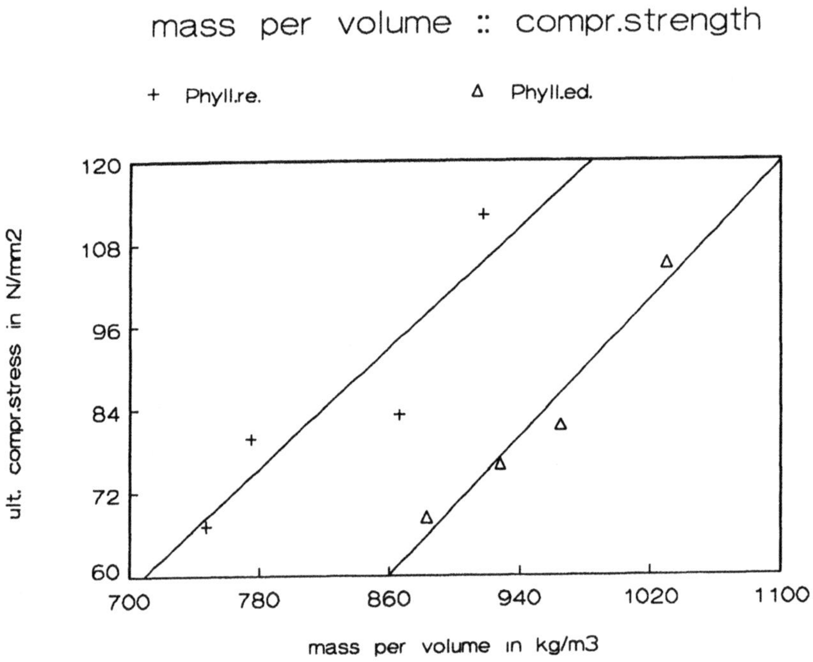

COMPRESSION 9

Keyword: Compression, mass per volume.
Subject: The ultimate compression stress related to the mass per volume.
Source: Zhou Fangchun, Studies on physical and mechanical properties of bamboo wood. Journal of Nanjing Techn. Coll. of For. Prod., 1981 no 2 pp 1-32. In Chinese; the following is based on an English translation of the headings of the tables and the captions of the figures only. Cited here are tables 20 and 21 and figure 19.

Results: From his table 21: (data related to the region of origin)

region	rho	σ'	n
Yixing	612	70.5	370
Xiashu	596	63.9	289
Shimen	589	59.9	233
Damaoshan	584	59.9	480

in which:
rho = mass per volume in kg/m^3,
σ' = ultimate compression stress in N/mm^2,
n = number of tests.
Formula: $\sigma' = 0.107*rho$, variation coeff. V = 17%.

Results from his table 20: (data related to the site class of the forest).

site class	rho	σ'	n
good	591	61.8	288
good/medium	597	64.7	140
medium	603	63.2	376
bad	602	65.8	299

Formula: $\sigma' = 0.107*rho$, variation coeff. V however is 44%.
From his figure 19 follows: $\sigma' = 0.153*rho - 34.5$.

COMPRESSION 10

Keyword: Moisture content, mass per volume, compression.
Subject: The influence of the moisture content and the mass per volume on the ultimate compression stress, for small specimens of bamboo from different parts of the culmwall.
Source: Ota, Motoi.
Studies on the properties of bamboo stem (part 9). On the relation between compressive strength parallel to grain and moisture content of bamboo splint.
Bull. Kyushu Univ. For., vol 22 (1953), pp 87-108.

Results: Motoi Ota has done compression tests on two species: Moso-chiku (Phyll.pubescens) and Ma-dake (Phyll.edulis). Compression tests have been done on small specimens, 348 in all, divided into four groups:

Type 1, with the complete cross-section of the wall of the culm,
Type 2, with the inner layer shaved off,
Type 3, with the outer layer shaved off, and
Type 4, with inner and outer layer shaved off.

Nine classes of moisture content have been used, see below. Results have been published by him as the minimum and maximum value out of 5 or 3 test results. In the next table, the present author presents the mean values of these minima and maxima (see next page).

Class of MC in %.	type 1 rho	type 1 σ'	type 2 rho	type 2 σ'	type 3 rho	type 3 σ'	type 4 rho	type 4 σ'
Phyll.pub.								
50-99	1125	67	1126	69	1153	62	1134	65
17-28	857	70	851	73	863	71	807	69
14-17	852	71	840	76	838	69	826	72
10-13	894	82	853	82	861	77	841	81
6- 8	858	110	840	102	851	105	831	101
5- 7	853	108	829	104	842	103	824	98
2- 4	843	129	823	123	831	128	813	117
0.7-2.0	833	143	816	142	819	141	805	133
0.3-1.0	830	147	809	137	817	137	804	135
Phyll.edulis.								
50-99	1168	64	1194	67	1190	63	1203	63
17-28	880	67	920	69	914	67	943	72
14-17	892	76	922	74	969	74	917	76
10-13	904	77	937	81	952	78	931	77
6- 8	890	102	929	108	931	100	918	100
5- 7	881	109	930	115	937	112	902	108
2- 4	874	119	921	124	933	120	914	111
0.7-2.0	874	135	909	141	912	141	884	132
0.3-1.0	869	140	911	144	910	131	902	132

in which rho = mass per volume in kg/m^3,
σ' = ultimate compression stress in N/mm^2.
In the opinion of the present author the differences between the four types are surprisingly small; presumably very thin layers have been shaved off.

If the MC is higher than the fibre saturation point, the stress is nearly a constant; if the MC is lower, the stress increases with decreasing MC.
For a MC lower than the fibre saturation point the next formula has been derived:

$$\sigma' = 147 * e^{(-0.045 * MC)}$$

with a correlation coefficient of 0.98.
The variation coefficient of the value -0.045 is 12.7%.
The fibre saturation point is 17.2%, with variation coefficient 10%.
Motoi Ota has determined the ratio between the compression strength of ovendry bamboo and water-soaked bamboo: the mean value is 2.19, with a variation coefficient of 7%.

He also has calculated the relationship between on one hand the ratio of the ultimate stress with the mass per volume and on the other hand the moisture content:

species and type	above fibre saturation point $\sigma'/rho = A + B*MC$			below fibre saturation point $\sigma'/rho = C*e^{\wedge}D*MC$		
	corr.c.	A	B	corr.c.	C	D
Phyll.pu						
1	-0.987	0.091	-0.00049	-0.978	0.174	-0.048
2	-0.967	0.090	-0.00046	-0.972	0.174	-0.052
3	-0.961	0.085	-0.00045	-0.974	0.176	-0.055
4	-0.864	0.090	-0.00047	-0.960	0.175	-0.055
Phyll.ed						
1	-0.989	0.081	-0.00038	-0.969	0.163	-0.046
2	-0.958	0.089	-0.00049	-0.982	0.166	-0.050
3	-0.973	0.083	-0.00045	-0.962	0.159	-0.049
4	-0.952	0.086	-0.00051	-0.951	0.155	-0.048

in which corr.c. = correlation coefficient.

Examples: for Phyll.pubescens, type 1, MC = 60% and mass per volume is 900 kg/m^3, the σ' = rho*(A + B*MC) =

$$900*(0.091 - 0.00049*60) = 79 \text{ N/mm}^2.$$

If the MC is 5% then: σ'/rho = 0.174*e^(-0.048*5) = 0.1369, from which follows σ' = 123 N/mm^2.

Finally, the properties adjusted to 12% MC are as follows.

species and type	rho kg/m^3	σ' N/mm^2	rho/σ'
Phyll.pub			
1	909	88.7	0.0976
2	912	85.2	0.0934
3	943	86.2	0.0914
4	888	80.7	0.0909
Phyll.ed.			
1	896	83.4	0.0931
2	942	85.5	0.0908
3	977	86.0	0.0880
4	968	84.1	0.0869

COMPRESSION 11

Keyword: Compression, position.

Subject: The ultimate compression stress related to the position in the culm.

Source: Zhou Fangchun, Studies on physical and mechanical properties of bamboo wood. Journal of Nanjing Techn. Coll. of For. Prod., 1981 no 2 pp 1-32. In Chinese; the following is based on an English translation of the headings of the tables and the captions of the figures only. Cited here is table 17 and figure 7.

Results: The author has determined the ultimate compression stress as a function of the relative height H of the culm; 0 means the bottom and 10 means the top. Results in N/mm²:

part of the culm H	0	2	4	6	8	n
σ from tests:						
region Xiashu	60.95	69.14	73.95	79.08	82.28	70
region Shimen	52.57	64.51	67.47	74 38	77.68	100

n = the number of tests. Regression formulae: Xiashu: $\sigma' = 61.19 + 4*H - 0.1713*H^2$; Shimen: $\sigma' = 53.42 + 4.9*H - 0.2366*H^2$.
The plot shows the stresses from the tests.

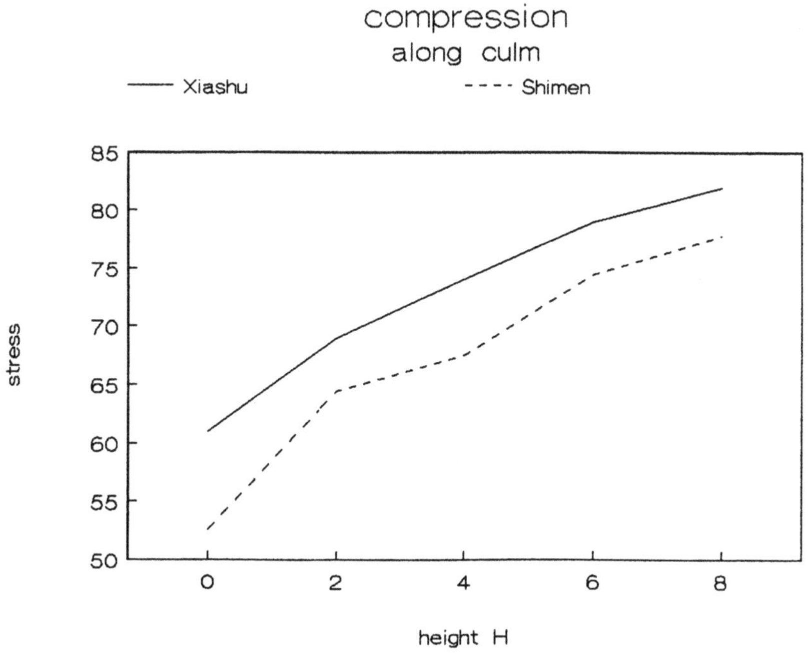

COMPRESSION 12

Keyword: Position, mass per volume and compression.
Subject: The influence of the position of the specimen in the internode on the mass per volume and the ultimate compression stress in split bamboo.
Source: Ota, Motoi.
Studies on the properties of bamboo stem (part 8). The variation of the specific gravity and the compressive strength of bamboo splint in a node-interval.
Bull. Kyushu Univ. For. (1953), vol 21, pp 71-82.

Results: Motoi Ota has studied two species, Phyll. ret. and Phyll. edulis. From each five cylinders have been selected, each consisting out of an internode with a node at each end, and each of them with a different thickness. From each cylinder specimens have been made crosswise: horizontal, i.e. in a circle around the culm just in the middle of the internode, and vertical, i.e. running in a straight line from one node to the other. Results from the tests in the horizontal direction are as follows (Each line represents presumably ten specimens).

no	thickness mm	MC %	rho kg/m^3	σ' N/mm^2
Phyll. ret.				
1	4.60-5.45	13.8-14.4	899-959	80-89
2	4.40-5.20	14.3-14.6	904-982	80-89
3	4.15-4.85	13.8-15.0	923-967	87-96
4	3.65-4.20	13.9-14.4	928-967	88-99
5	3.45-3.95	13.1-13.8	921-955	87-95
Phyll.ed.				
1	11.25-13.75	18.8-20.0	713-793	64-79
2	9.80-12.25	18.9-20.6	784-852	70-89
3	8.20-9.05	17.2-19.0	845-902	76-87
4	5.45-6.00	16.3-17.8	942-966	91-97
5	4.15-5.25	16.0-18.1	933-1009	90-108

in which no = number of the cylinder.

The thickness of the culm is rather constant in the central 70% of the length of the internode, and increases in the remaining 15% above and below the node. The variation of the MC along the length of the internode is rather much. As to the mass per volume (rho), for Ph.ret. the variation along the internode is rather much, and the rho decreases towards the node, from between 900 and 1000 kg/m³ to about 850; for Ph. ed. the rho is rather constant along the internode, and increases a little towards the nodes. The compressive strength varies along the internode, and decreases towards the nodes. The foregoing conclusions summarize the figures 2 to 5 in the original article.

The coefficients of variation are as follows.

	mass per volume				ultimate compr stress			
	Ph.ret.		Ph.ed.		Ph.ret.		Ph.ed.	
no	hor.	vert.	hor.	vert.	hor.	vert.	hor.	vert.
1	2.17	0.78	4.26	0.92	4.32	4.48	6.80	4.26
2	2.11	0.51	2.78	0.69	3.42	4.29	7.74	2.05
3	1.28	0.99	2.04	1.05	3.17	3.37	3.56	3.13
4	1.15	0.71	1.01	0.84	3.12	3.86	2.12	3.55
5	1.10	0.64	3.06	2.86	2.55	2.55	6.55	6.76

The coefficients of variation for the vertical direction have been taken from the central 80% of the internode.

COMPRESSION 13

Keyword: Compression, age.
Subject: The ultimate compression stress, related to the age at the time of cutting.
Source: Zhou Fangchun, Studies on physical and mechanical properties of bamboo wood. Journal of Nanjing Techn. Coll. of For. Prod., 1981 no 2 pp 1-32. In Chinese; the following is based on an English translation of the headings of the tables and the captions of the figures only. Cited here is table 10 and figure 6.

Results: The ultimate compression stress in N/mm^2 is as follows.

age class A	1	2	3	4	5	
age in years	1-2	3-4	5-6	7-8	9-10	n
district:						
Xiashu	53.74	66.10	67.15	70.06	62.52	289
Yixing	65.94	72.01	72.68	71.74	70.30	370
Shimen	54.42	57.71	64.00	65.10	58.54	233
Damaoshan	48.30	59.12	61.44	66.53	64.26	480
mean	55.60	63.74	66.32	68.36	63.91	1372

in which n = number of tests. A regression formula for the mean is: $\sigma' = 44.4 + 13.1*A - 1.83*A^2$. The best age is 7.2 years.

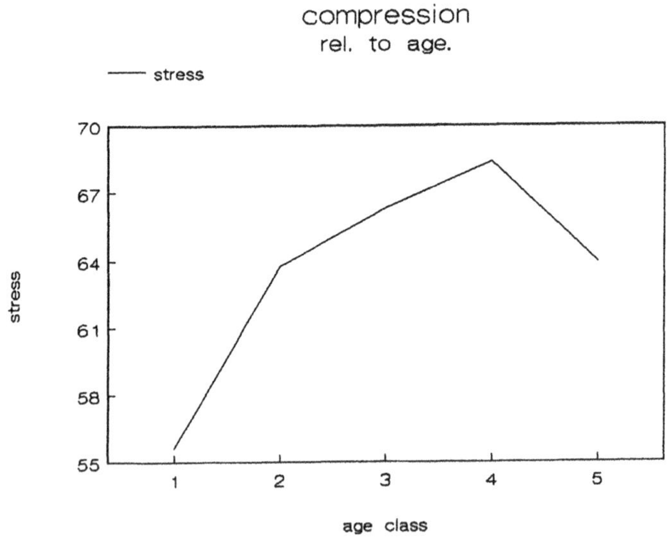

COMPRESSION 14

Keyword: Compression, felling season.
Subject: The ultimate compression stress related to the felling season, the months of the year.
Source: Zhou Fangchun, Studies on physical and mechanical properties of bamboo wood. Journal of Nanjing Techn. Coll. of For. Prod., 1981 no 2 pp 1-32. In Chinese; the following is based on an English translation of the headings of the tables and the captions of the figures only. Cited here is figure 10.

Results: The next diagram shows the ultimate compression stress in N/mm^2 as a function of the felling season, i.e. the months of the year. This graph is valid for bamboo from Shimen; for each region and species a similar relationship can be determined.

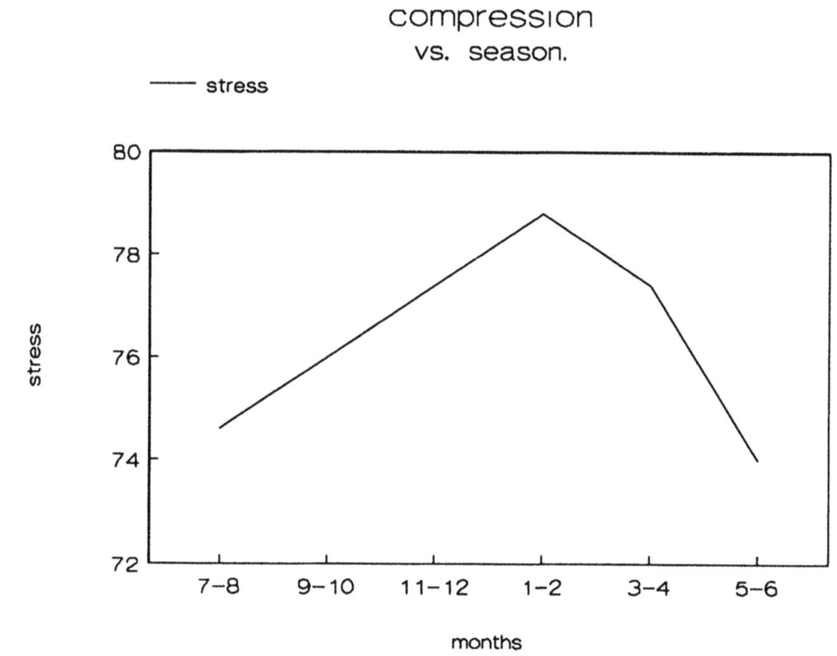

COMPRESSION 15

Keyword: Compression, age, position, mass per volume.
Subject: Compression related to age, position in the culm and mass per volume.
Source: Limaye, V.D. Strength of bamboo (Dendr. strictus). Indian Forest Records, new series, timber mech., vol 1 no 1 (1952), 17 pp; reprinted from The Indian Forester, vol 78 no 11 (1952), pp 558-575.

Results: Limaye has collected a 200 culms, 50 of each age (0.5 1, 2 and 2.5 years), and tested them on bending and compression in green and dry condition. The diameter was between 25 and 62 mm, and the length of the culms was between 5.4 and 7.8 m. According to diameter, wallthickness, length of culm and length of internode they were matched into pairs, and one of each pair was tested green and the other one dry.
The results of the tests on compression are as follows.

		green						dry					
		N in C			C inter N			N in C			C inter N		
age	P	n	MC	σ'	n	MC	σ'	n	MC	σ'	n	MC	σ'
0.5	B	11	112	21.7	11	111	23.4	11	8.9	51.7	11	9.0	56.2
	M	11	92	23.8	11	102	23.4	11	8.9	54.8	11	8.8	53.7
	T	11	87	24.8	11	104	23.4	11	9.1	56.2	11	8.4	53.7
	m	33	97	23.4	33	106	23.4	33	9.0	54.4	33	8.8	54.4
									12.0	43.1		12.0	42.7
1	B	9	73	25.1	9	89	25.5	9	12.7	45.1	9	12.4	45.1
	M	9	79	25.1	9	95	25.1	9	12.0	47.5	9	11.8	48.6
	T	9	70	24.8	9	95	25.5	9	12.3	45.8	9	11.8	44.4
	m	27	74	25.1	27	93	25.5	27	12.3	46.2	27	12.0	46.2
									12.0	46.2		12.0	46.2
2	B	10	60	36.2	10	66	34.5	10	9.3	57.9	10	10.5	54.8
	M	10	52	35.5	10	57	36.5	10	9.1	54.4	10	10.4	58.2
	T	10	44	35.1	10	51	36.2	10	8.2	51.3	10	12.6	53.1
	m	30	52	35.5	30	58	35.5	30	8.9	54.8	30	11.2	55.1
									12.0	48.6		12.0	53.1
2.5	B	11	73	40.3	11	62	38.9	11	8.9	66.5	11	9.2	68.9
	M	11	61	45.1	11	62	39.3	11	8.8	71.7	11	9.0	74.4
	T	11	58	44.4	11	60	38.9	11	8.7	70.6	11	9.0	74.8
	m	33	64	43.4	33	61	38.9	33	8.8	69.6	33	9.0	72.7
									12.0	60.6		12.0	61.3

in which N and C mean node and centre resp.,
age = age in years,
P = position along the culm, B=bottom, M=middle, T=top, m=mean
n = number of tests
MC = moisture content
σ' = ultimate compression stress in N/mm^2.

The statistical calculations, done by Limaye, show the following results. *Firstly*, an analysis of variances.

	DF	σ'
B = between ages	3	1244 "
C = centre inter node vs. node inter c.	1	14.7
B*C	3	20.1
error (1)	-	20.7
(DF =)		(74)
D = between positions in the culm	2	8.3
B*D	6	2.8
C*D	2	2.4
B*C*D	6	5.2
error (2)	-	2.5
(DF =)		(141)

in which
DF = degree of freedom,
σ' = ultimate compr.stress in N/mm²,
" = significant at 0.1 percent level,
' = idem at 5 percent.

Secondly, the difference between node in centre or centre inter node:

node in centre : σ' = 31.7 N/mm²,
centre inter node: σ' = 31.0,
standard error : 0.8.

Thirdly, the position along the culm:

bottom: σ' = 30.7 N/mm²,
middle: 31.7
top : 31.7
st.error: 0.34.

Both influences taken together:

	bottom	middle	top	
node in centre	σ'= 30.7	32.4	32.4	N/mm²,
centre inter node	30.7	31.0	31.0	
standard error for B+M+T = 0.86.				

Finally, the relationship between σ' and the age for green bamboo is:

age 0.5 year:	σ' = 23.4 N/mm²,
1	25.1
2	35.5
2.5	41.3
standard error	1.1

from which one can calculate: $\sigma' = 17.5 + 9.2*\text{age}$, with $R = 0.99$.

The improvement of σ' with age is for kilndry bamboo: (the MC has been adjusted to 12 percent)

age in years	0.5	1	2	2.5
σ' in N/mm²	43.1	46.2	50.6	61.0

COMPRESSION 16

Keyword: Age, moisture content, position, compression.
Subject: Compression tests related to the age at the time of cutting, the moisture content and the position in the culm.
Source: Sekhar, A.C., B.S. Rawat and R.K. Bhartari.
Strength of bamboos: Bambusa Nutans.
Indian Forester (1962, January) pp 67-73.

Results: In a similar way as they have organized their tests on bending, the authors have carried out tests on compression. The results are as follows.

			green		dry	
a	P	n	MC	σ'	n	σ'
1	B	8	125	35.2	8	64.8
	M	8	121	35.4	7	63.5
	T	8	116	38.0	8	66.5
	m	24	121	36.2	23	64.9
2	B	8	91	43.8	8	64.2
	M	8	88	45.7	8	74.8
	T	8	76	54.8	8	69.7
	m	24	85	47.1	24	69.5
3	B	8	98	43.7	8	87.9
	M	8	83	44.8	8	83.9
	T	8	79	48.9	8	85.6
	m	24	87	45.8	24	85.4
4	B	7	81	49.4	7	53.5
	M	7	90	54.2	7	59.4
	T	7	84	45.1	7	61.3
	m	21	85	48.6	21	58.1
5	B	8	80	47.3	8	69.2
	M	8	74	47.6	8	71.8
	T	8	77	50.2	8	64.1
	m	24	77	48.4	24	68.4

in which:
a = age in years,
P = position along the culm,
B/M/T = bottom/middle/top,
m = mean,
n = number of tests,
MC = moisture content,
σ' = ultimate compression stress in N/mm^2.

A relationship between the ultimate stress in compression and the age cannot be found, which is surprising.
In their figure 1, the original authors mention a ratio between the mass per volume and the ultimate stress (for green condition):

$$\sigma = 0.003 * (rho)^{1.5}$$

in which σ is in N/mm^2 and rho in kg/m^3. The present author has done similar calculations with a slightly different result:

$\sigma = 0.00897 * (rho)^{1.33}$ with $R = 0.90$ (R = square root of coeff. of determination). For dry condition a good fit could not be found.

Tests on static and impact bending and on shear have been done on specimens from the same culms.

COMPRESSION 17

Keyword: Age, mass per volume and compression.
Subject: The mass per volume and compression strength related to the age.
Source: Sekhar, A.C. and R.K. Bhartari. A note on the strength of dry bamboo (Dendrocalamus strictus) from Madhya Pradesh. Indian Forester (1961, Oct.) pp 611-613.

Results: Material was collected from two localities in the said region. Pieces were about 3.30 m long, and in each age group 20 pieces were collected. Clear specimens only have been tested, conditioned to about 12% MC. Results:

age year	rho kg/m^3	n	σ' N/mm^2
1	646	14	51.8
2	703	17	55.3
3	718	16	54.4
4	706	17	58.3
5	672	18	58.2
6	608	16	46.3

in which:
age = the age in years at the time of cutting,
rho = the mass per volume in kg/m^3,
n = the number of tests,
σ' = the ultimate compression stress in N/mm^2.

A relationship between σ' and rho or age cannot be found.

On the same specimens tests have been done on the size of the culm, static and impact bending.

COMPRESSION 18

Keyword: Age, compression.
Subject: The ultimate compression stress in Phyll.glauca cut at ages from 1 to 7 years.
Source: Lu Xiu-xin, Wang Ke-qing, Yi Xiang-chen, Liou Jin and He Ji-xun. A study on the physico-mechanical properties of culmwood of Phyll.glauca of Shandong. J. of bamboo res., Zhejiang For. Inst., Hangzhou, China, vol 4 no 2 (July 1985) pp 98-106.

Results: The authors have done tests on 54 culms from 4 different regions in the Shandong province; age classes from 1 to 7 years, and two culms per region and yearclass. Tests have been done on moisture content, mass per volume, tension, compression and bending, to determine the influences of the age at time of cutting and the region of origin. The number of tests was over 200 for each test and each region. The results dealing with the ultimate compression stress have been done on specimens of a size of 20 mm height, 20 mm width, and wallthickness. Results:

region age	Sancha	Dajinkou	Dahuaya	Luchanya
1 year	61.0	41.2	58.7	56.8 N/mm²
2	71.8	49.5	57.6	64.2
3	70.7	58.1	63.9	71.4
4	73.7	67.6	69.5	88.0
5	72.0	67.3	76.4	86.3
6	71.5	64.7	70.0	73.9
7	75.6	63.8	68.4	

The next polynomials have been calculated (a = age in years).

Region	Ult.compr.stress in N/mm^2	st.dev.	best age
Sancha	$\sigma' = 58.8 + 5.39*a - 0.48*a^2$	3.68	5.67
Dajinkou	$26.2 + 15.4*a - 1.45*a^2$	2.43	5.33
Dahuaya	$47.7 + 8.47*a - 0.76*a^2$	4.64	5.56
Luchanya	$32.9 + 22.56*a - 2.53*a^2$	7.06	4.45

COMPRESSION 19

Keyword: Compression.
Subject: Compression tests on split bamboos.
Source: Ueda, K. Mechanical properties of Moso bamboo (Phyll.pub.). In: Res. Bull. Coll. Exp. For. Hokkaido Univ., vol 37 no 3, (1980), pp 817-836.

Results: The author has done tests on bending, compression and torsion on 51 specimens of Phyll.pub. The tests on compression have been carried out on composed specimens: four pieces of split bamboo of a width of 10 mm, a thickness of 10 mm and a length of 50 mm have been glued together with the soft innersides inside to a sample of 20 mm wide, 20 mm thick and 50 mm long. Results are as follows.

No. of specimen	rho kg/m^3	σ.el N/mm^2	σ' N/mm^2	σ.el/σ'
01	740	46.6	76.8	0.61
04	740	39.6	74.1	0.53
05	720	32.0	70.8	0.45
06	730	31.0	71.7	0.43
29	740	25.5	72.4	0.35

COMPRESSION, TESTMETHOD 1

Keyword: Compression, testmethod.
Subject: Compression tests with and without lead plates.
Source: Meyer, H.F. and B. Ekelund.
Tests on the mechanical properties of bamboo.
The Engineering Soc. of China. Session 1922-1923, vol 22 paper 7, pp 141-169. Present in library E.T.H. Switzerland.

Results: Tests have been carried out in two different ways. In three tests the ends of the specimens were covered with a lead plate of 2 mm thick, resulting in an ultimate compression stress of 34.5 N/mm^2, with a standard deviation of 1.6 N/mm^2. Four tests have been done without steel plates, with an ultimate compression stress of 44.2 N/mm^2 and a standard deviation of 5.2 N/mm^2. It is interesting to see that already in 1922 the phenomenon of contraction has been studied.

COMPRESSION, TESTMETHOD 2

Keyword: Compression, test method.
Subject: The influence of the size (height and width) of the specimen on the ultimate compression stress in split bamboo.
Source: Ota, Motoi.
Studies on the properties of bamboo stem (part 4). Study on the form of specimen in the test of compressive strength.
J. Jap. For. Soc. (1950), vol 32, pp 65-69.

Results: Motoi Ota has carried out many tests on the ultimate compressive stress for split bamboo specimens for the species Phyll. reticulata, Ph. pubescens and Ph. nigra. If we call:

t = the wall thickness of the bamboo culm,
b = the width of the split specimen (tangential), and
h = the height of the specimen,running parallel with the length of the culm,

then he has performed compression tests on many specimens with ratios of h/t from 1 to 5 (or 7 or even 11) and b/t from 0.7 to 4. His conclusion is, that the ultimate compression stress is not influenced by the ratio b/t, but decreases with increasing h/t, roughly as follows:

$$\sigma = -90 + 2.5*(h/t)$$

in which σ = ultimate compression stress in N/mm^2; h = height of the split specimen, t = wallthickness.
From this formula there is no optimum to be found, but the original author states:"The appearance rate of normal type of failure.....is maximum at b/t = 1.0 to 1.5 and h/t = 2.0 to 2.5." Based on this conclusion he advises to do compression tests on split specimens with ratios b/t = 1 and h/t = 2.

COMPRESSION, TESTMETHOD 3

Keyword: Compression, test methods.
Subject: The ultimate compression stress of a full culm, compared with the same of parts of a culm.
Source: Zhou Fangchun, Studies on physical and mechanical properties of bamboo wood. Journal of Nanjing Techn. Coll. of For. Prod., 1981 no 2 pp 1-32. In Chinese; the following is based on an English translation of the headings of the tables and the captions of the figures only. Cited here are tables 5, 6 and 7.

Results: The author has done compression tests on full culms, and on split parts of culms. He studies the influence of this form on the ultimate compression stress. Results in N/mm^2:

crosssection	full	1/2	1/4	1/8	1/16	1/32
σ', mean m	64.96	58.52	58.67	58.53	59.99	57.36
standard dev., s	3.06	3.53	5.25	7.95	5.33	5.63

The number of tests was 9 in each case, 54 in total.

In a first analysis of variances, with 5 degrees of freedom for the cut sections, and 48 degrees of freedom for the specimens, he calculates F = 2.32 (in a Fisher-distribution). The critical values are 2.0 at a 10%-level and 2.41 at a 5%-level. Conclusion: the difference between a compression test on a full cross-section of a culm and a split part of it, is significant at a 10-percent level, not at a 5 percent level.
In a second analysis of variances, with 4 degrees of freedom for the cut sections (also without the full culm) and 40 for the specimens, is shown a F of 0.241; F critical is in this case 2.61 at a 5%-level. The differences between the cut sections are not significant.

COMPRESSION, TESTMETHOD 4

Keyword: Compression, test methods.
Subject: The ultimate compression stress as a function of the number of nodes in the specimen.
Source: Atrops, J.L. Elastizitaet und Festigkeit von Bambusrohren (elasticity and strength of bamboo culms). Der Bauingenieur (1969) vol 44 no 6 pp 220-225. In German.

Results: Atrops has studied the ultimate compression stress for specimens without a node, with 1 node and with 2 nodes. Results:

number of nodes	form of specimen		d mm	t mm	σ' N/mm²
0	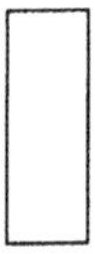	min.	71	5.6	38.6
		mean	77	6.9	40.2
		max.	94	11.0	52.2
1	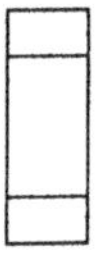	min.	72	5.6	36.2
		mean	76	7.0	40.7
		max.	92	11.1	53.7
2	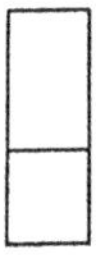	min.	72	5.7	35.6
		mean	77	6.9	43.3
		max.	93	11.1	52.8

in which d =diameter in mm; t = wallthickness in mm; σ' = ultimate compression stress in N/mm².
The height of the specimens was 4*d,
moisture content MC = 18.1 percent,
speed of testing 0.33 N/mm²/s,
number of tests 36 each, 108 in total.

The author mentions also tests on small specimens, wide 16 mm, thick t (= wallthickness), and high 25 mm, with σ' = 62.1 N/mm². The E-modulus is between 16500 and 21600 N/mm².

COMPRESSION, PERPENDICULAR

Keyword: Compression.
Subject: The compression strength of a culm, loaded in a direction perpendicular to the axis.
Source: Atrops, J.L. Elastizitaet und Festigkeit von Bambusrohren (elasticity and strength of bamboo culms). Der Bauingenieur (1969) vol 44 no 6 pp 220-225. In German.

Results: Atrops has carried out compression tests on specimens of full bamboo culms, length presumably 400 mm, diameter 65 to 85 mm, and wallthickness 6 to 10 mm. These specimens have been loaded with a force perpendicular to the axis, with an intermediate steel plate of 100 mm square. Four tests series have been carried out.

1. Without nodes, mean strength F = 5.25 kN,
2. with 2 nodes, F = 6.35 kN,
3. with 1 node, F = 9.30 kN.

The number of tests was 24 in each case.

Side views:

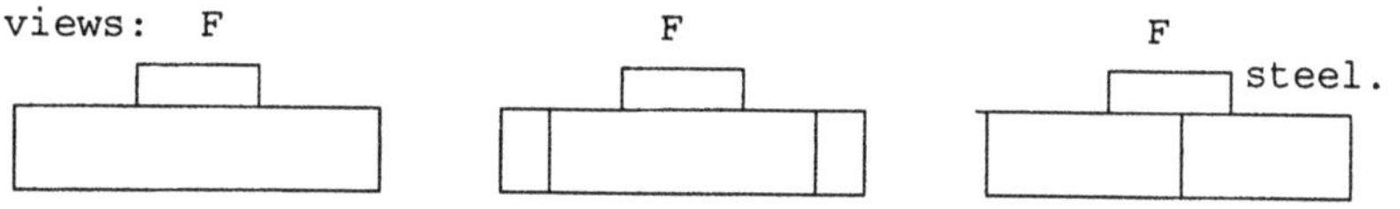

bamboo without nodes, with two nodes, and with one node.

Evidently the open piece, without nodes, is the weakest, and the bamboo with one node is the strongest: this node supports the force F. The specimen with two nodes is in between.

4.The last test series deals with two bamboo specimens, without nodes and without steel plates, perpendicular to another, loaded with a force F perpendicular to their axes, F = 5.80 kN.

BUCKLING 1

Keyword: Buckling.
Subject: Buckling tests.
Source: Meyer, H.F. and B. Ekelund.
Tests on the mechanical properties of bamboo.
The Engineering Soc. of China. Session 1922-1923, vol 22 paper 7, pp 141-169. Present in library E.T.H. Switzerland.

Results: It is really remarkable to read about buckling tests already performed as early as 1922. Tests have been done on bamboos of 1.2, 1.5, 1.8 and 2.1 metres. The outside diameter was between 30 and 70 mm.The original data give the ultimate load, the buckling length, the outside and inside diameter. From these data the present author has calculated the next buckling diagram, resulting in the formula:

y = 44.8 * e ^(-0.0089 * x)

in which:

y = the ultimate buckling stress in N/mm^2,
x = the slenderness = lambda.

The square root of the coefficient of determination R = 0.80.
The diagram is:

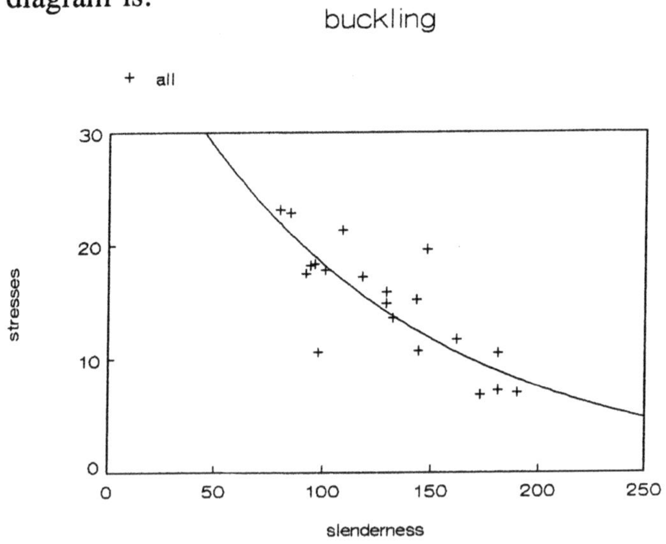

BUCKLING 2

Keyword: Buckling.
Subject: The ultimate buckling stress σ'as a function of the slenderness lambda.
Source: Atrops, J.L. Elastizitaet und Festigkeit von Bambusrohren (elasticity and strength of bamboo culms). Der Bauingenieur (1969) vol 44 no 6 pp 220-225. In German.

Results: Buckling tests on bamboo are hardly heard of. One of the exceptions is the report by Atrops, who did buckling tests as early as 1969 in Trinidad. Unfortunately he does not give a full report on the procedure. He presents his results in the next plot.
Horizontal axis: the slenderness lambda,
vertical axis: the ultimate buckling stress in kp/cm^2 = 10 N/mm^2, from which we can derive a range from 10 to 45 N/mm^2.
The upper curve represents culms with a diameter/wallthickness or D/T ratio of 6; the lower is valid for D/T = 10.

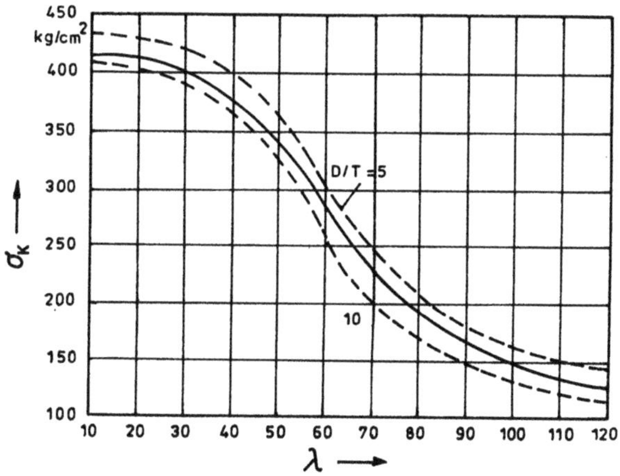

8. Shear

SHEAR 1

Keyword: Shear.
Subject: Shear tests on green and dry bamboos; node and internode.
Source: Meyer, H.F. and B. Ekelund. Tests on the mechanical properties of bamboo. The Engineering Soc. of China. Session 1922-1923, vol 22 paper 7, pp 141-169. Present in library E.T.H. Switzerland.

Results: To carry out shear tests, pieces of full bamboo culm have been provided with four square holes as shown in the figure, copied from the original article. By applying load at the bottom, the iron bar was pushed upwards. The results are as follows.

	Ultimate shear stress σv in N/mm^2.			
	internode		node	
	seasoned	green	seasoned	green
no of tests	5	5	8	2
σv, mean	8.64	7.42	8.56	8.10
st.dev.	1.22	0.74	1.05	--

The figure from the original article is as follows.

SKETCH OF SHEARING TEST

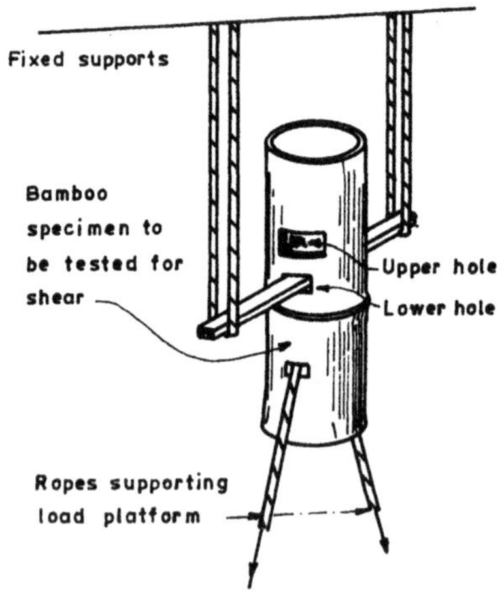

SHEAR 2

Keyword: Moisture content, shear, mass per volume.
Subject: The ultimate shear stress for Phyll.pubescens, Ph. retic. and Ph. nigra, for a MC from dry to soaked.
Source: Ota, Motoi. Studies on the properties of bamboo stem (part 13). On the relation between shearing strength parallel to grain and moisture content of bamboo splint.
Bull. Kyushu Univ. For. (1955), vol 25, pp 121-131.

Results: Motoi Ota has tested 112 specimens on shear, divided into 7 classes according to their M.C., see next table.

treatment	class	moisture content in %					
		Ph.pub.		Ph.ret.		Ph.nigra	
		m	V	m	V	m	V
70	1	2.4	12.42	2.1	4.86	2.0	5.83
50	2	7.0	3.17	6.7	0.73	6.4	1.25
40	3	10.2	2.04	10.0	1.44	9.1	5.20
30	4	14.6	1.65	14.1	3.61	14.1	2.52
20	5	16.9	3.70	15.9	1.17	15.5	0.75
0	6	35.1	25.35	32.0	17.48	22.5	7.29
watersoaked	7	102.4	18.58	87.6	15.96	78.9	4.69

in which:
treatment = density of H_2SO_4 in the desiccator in %.
m = mean of the sample,
V = coefficient of variation,
number of specimens = 112 in all.

In the case of class 6, the lower part of the desiccator was filled with water. This caused fungi, resulting in a high coefficient of variation. For this reason the results from the specimens of class 6 were treated as reference data only.
Data on the mass per volume (rho) of the specimens are as follows (adjusted to 15% M.C.) (V = coeff. of variation, and
t = wallthickness in mm).

species	specimen	t (mm)	rho (kg/m^3)	V in %
Ph.ret.	7- 4	16.25	603	1.81
	5-18	9.10	823	2.04
	7-12	10.35	721	0.73
	8-12	8.10	754	1.13
	7-16	9.95	750	1.35
	3-17	8.50	803	1.09
Ph.ed.	9-10	6.35	799	1.43
	5-10	5.30	701	1.87
	6-12	5.05	718	1.61
	8- 8	6.00	857	0.64
	7-20	4.45	916	1.23
Ph.nigra	7-14	5.40	840	1.61
	5-14	4.10	856	0.49
	2-10	6.25	822	1.47
	1-16	5.15	790	1.92
	4-14	3.80	812	1.78

The results of the shear tests are as follows.

class		1	2	3	4	5	6	7
min. MC in %		1.8	6.3	8.4	13.2	15.4	20.6	68.3
max. MC in %		2.6	7.4	10.5	14.9	17.9	48.2	140.7
species	specimen	ultimate shear stress in N/mm²						
Ph.ret.	7- 4	25.8	17.9	13.7	12.4	11.5	10.3	11.2
	5-18	27.3	29.7	24.2	21.8	20.4	16.0	15.9
	7-12	17.9	19.7	17.2	14.7	14.3	13.1	13.4
	8-12	19.4	22.0	23.3	18.8	17.8	14.2	13.3
	7-16	20.4	19.4	20.5	17.1	15.0	10.9	11.7
	3-17	28.6	23.0	24.9	19.4	17.9	16.6	17.8
Ph.ed.	9-10	24.2	20.8	18.5	16.7	16.4	9.6	10.1
	5-10	15.6	20.8	15.6	15.7	12.7	10.3	9.7
	6-12	17.5	18.1	18.2	15.3	12.6	8.8	8.6
	8- 8	25.6	27.9	23.8	22.2	19.3	8.3	14.3
	7-20	24.3	24.8	23.4	21.1	19.0	11.2	17.4
Ph.nigra	7-14	24.6	24.9	24.6	20.2	17.9	11.4	10.8
	5-14	20.3	24.1	23.3	20.0	20.5	16.9	16.7
	2-10	26.3	25.0	20.3	16.6	15.3	15.8	15.3
	1-16	21.1	25.0	23.9	18.0	15.7	12.7	10.8
	4-14	24.2	22.3	22.1	19.3	15.8	11.1	13.7

From these test results Motoi Ota has derived the next data.

species, specimen	n	σv = a - b*MC a	b	c	f.s.p.	σv at 15% MC
Ph.ret.						
7-4	3	16.9	0.31	-0.999	18.7	12.3
5-18	4	35.5	0.93	-0.959	21.2	21.6
7-12	4	22.6	0.49	-0.966	18.6	15.2
8-12	3	32.2	0.91	-0.991	20.7	18.5
7-16	3	29.3	0.85	-0.999	20.6	16.5
3-17	3	36.0	1.11	-0.999	16.3	19.3
whole	20	29.4	0.81	-0.606		
Ph.ed.						
9-10	5	24.4	0.51	-0.954	27.8	16.7
5-10	4	23.2	0.59	-0.829	22.7	14.3
6-12	3	27.9	0.93	-0.960	20.8	14.0
8- 8	4	33.4	0.87	-0.971	22.0	20.4
7-20	3	30.4	0.71	-0.999	18.3	19.8
whole	19	27.7	0.71	-0.650		
Ph.nigra						
7-14	3	35.4	1.10	-0.987	22.2	18.8
5-14	4	26.9	0.44	-0.977	23.3	20.3
2-10	4	31.1	1.04	-0.988	15.3	15.5
1-16	4	32.1	1.02	-0.986	20.9	16.8
4-14	3	30.8	0.93	-0.971	18.5	16.9
whole	18	30.2	0.83	-0.882		
whole	57	29.0	0.78	-0.690	20.4	

in which n = number of tests
σv = ultimate shear stress in N/mm²
c is the corr. coeff. of: σv = a - b*MC, which formula is valid below the f.s.p. = fibre saturation point in %

SHEAR 3

Keyword: Shear, test method.
Subject: The influence of the form of the specimen in the shear test on the resulting shear stress.
Source: Ota, Motoi.
Studies on the properties of bamboo stem (part 12). On the form of stair type specimen in the shearing test parallel to the grain. Bull. Kyushu Univ. For., vol 25 (1955), pp 73-84.

Results: Motoi Ota has carefully studied the influence of the form of the specimen on the resulting shear stress. Firstly data on the test material are given.

	moisture content in %				rho in kg/m³ at 15% MC			
	Phyll.pub.		Phyll.ed.		Phyll.pub.		Phyll.ed.	
n	m	V	m	V	m	V	m	V
A 12	15.2	1.17	13.9	3.41	699	3.32	747	3.62
B 12	15.2	0.83	15.2	2.91	713	2.68	723	0.74
C 12	15.1	1.18	14.4	3.47	729	2.72	767	4.07
D 12	15.5	1.04	15.3	2.68	724	1.38	845	3.48
E 12	15.2	1.02	14.2	3.65	747	1.88	854	2.69
whole	15.2	1.21	14.6	4.76	722		787	

in which n = number of specimens
m = mean of the sample
V = coefficient of variation in %

Next, the type of the specimen is:

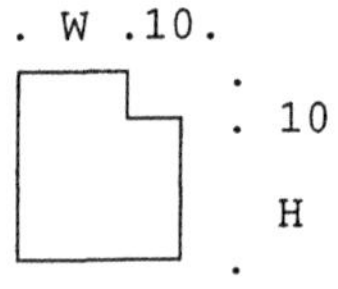

which means that the used stairtype specimen has a width of W + 10 mm, a heigth of H + 10 mm, and the thickness is the wallthickness. Variables are W (10,20 and 30 mm) and H (10,20,30 and 40 mm).

A summary of his test results is as follows.

W =	10				20				30			
H =	10	20	30	40	10	20	30	40	10	20	30	40
Phyll.pub.												
σ_v=	13.5	13.2	12.1	10.5	10.9	12.5	11.2	10.3	12.9	12.4	11.1	10.5
s =	3.1	1.7	1.1	0.7	3.0	1.3	1.6	0.4	2.1	1.2	1.0	1.6
Phyll.ed.												
σ_v=	13.2	16.2	13.6	13.0	13.5	14.9	13.0	13.5	12.9	15.4	14.5	11.7
s =	1.9	3.4	2.7	1.2	5.1	3.5	2.5	1.9	4.7	2.5	2.2	2.3

in which σ_v = shear stress in N/mm^2, mean value of 5 tests,
s = standard deviation in N/mm^2.

Motoi Ota argues that "the highest uniformity of shearing strength is obtainable from the H = 20 type specimens" as well as from the W = 20 type specimens. His conclusion is based on a study of the coefficients of variation; the present author is unable to understand this study to the full.

SHEAR 4

Keyword: Age, shear.
Subject: Shear tests related to the age at the time of cutting.
Source: Sekhar, A.C., B.S. Rawat and R.K. Bhartari.
Strength of bamboos: Bambusa Nutans.
Indian Forester (1962, January) pp 67-73.

Results: The authors have done shear tests on the bamboo culms on which also bending and compression tests have been reported. Tests on shear have been done in a similar way as Motoi Ota has developed, the stairtype specimen:

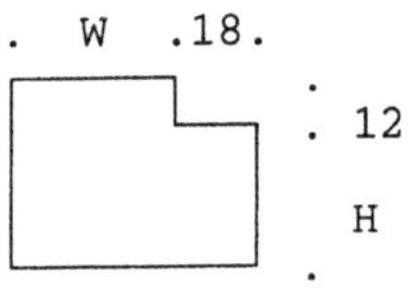

The width W varies between 12 and 50 mm, and the height H between 25 and 50 mm. The reported values for the ultimate shear stress are the mean values for all sizes, and all positions in the culm. All tests have been done on green bamboo. Results:

age	n	MC	σv
1	24	109	6.75
2	36	85	7.65
3	36	85	7.93
4	33	75	9.82
5	48	82	7.88

in which age is in years,
n = number of tests,
MC = moisture content in %,
σv = ultimate shear stress in N/mm^2.

Minimum shear stress occurs for W = 12 mm and H = 25 mm, and maximum shear stress for W = 50 and H = 38 mm.
It would have been very interesting if data had been reported more specifically.
Tests on static and impact bending and on compression have been done on specimens from the same culms.

SHEAR 5

Keyword: Shear, test method.
Subject: The ultimate shear stress as a function of the form of the specimen.
Source: Atrops, J.L. Elastizitaet und Festigkeit von Bambusrohren (elasticity and strength of bamboo culms). Der Bauingenieur (1969) vol 44 no 6 pp 220-225. In German.

Results: Atrops has done shear tests on specimens with a cross-section of a quarter of a culm, type A, and on specimens cut from a full culm, type B. Results in N/mm²:

specimen type	A	B
no. of tests	32	32
ult. shear stress:		
minimum	16.9	14.7
mean	19.8	16.7
maximum	23.1	22.2

The next figure shows the sizes and the form of the specimens. Sizes are in mm.

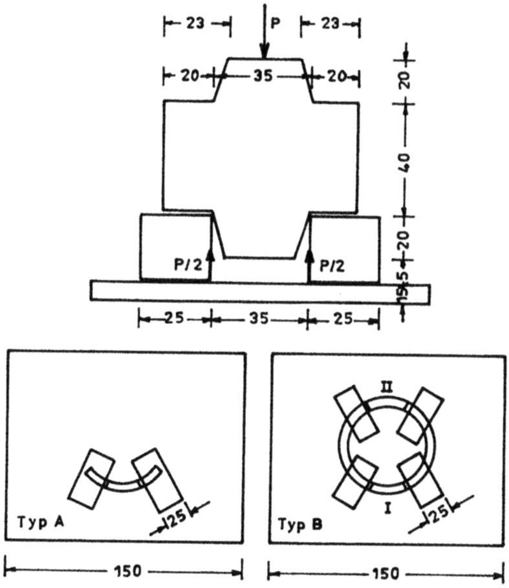

9. Tension

TENSION 1

Keyword: Moisture content, tension.
Subject: The tensile strength of two bamboo species related to moisture content.
Source: Ota, Motoi.
Studies on the properties of bamboo stem (part 10). On the relation between the tensile strength parallel to the grain and the moisture content of bamboo splint.
Bull. Kyushu Univ. For., vol 23 (1954), pp 155-164.

Results: Motoi Ota has tested bamboo specimens on tensile strength. Unfortunately the present author is unable to find a description of the form of the specimen used in the tests. Table 1 of the article presents the following results.

species and specimen.	thickness mm	rho kg/m³	MC %	σ' N/mm²	σ N/mm²
Phyll.ed.					
A	11	871	13.4	63.4	153.3
B	12.7	826	13.6	57.7	150.3
C	9.8	620	13.6	38.7	112.2
D	9.2	820	13.0	63.5	238.5
E	8.7	892	13.1	71.8	239.8
Phyll.pub.					
A	5.3	762	13.5	53.1	248.0
B	6.0	891	13.1	72.3	251.5
C	7.3	788	13.0	74.2	264.3
D	5.7	949	13.1	75.0	288.3
E	5.3	761	13.6	66.9	262.4

His table 2 gives tensile strength related to moisture content:

MC in %	2.0-3.8	2.4-3.9	8.0-10.3	13.5-15.3	17.9-23.5	54.2-86.9
Phyll.ed				σ in N/mm²		
A	122.2	121.6	146.6	153.3	136.5	143.8
B	101.7	139.7	152.8	150.3	191.8	142.9
C	91.0	97.7	134.6	112.2	150.0	65.4
D	141.9	132.9	163.1	238.5	204.3	225.0
E	134.7	135.9	170.3	239.8	200.4	238.0
Phyll.pub.						
A	125.1	135.5	180.0	248.0	126.1	177.6
B	168.6	232.7	179.4	251.5	146.9	218.4
C	191.2	225.5	253.0	264.3	169.4	189.3
D	329.6	276.8	297.7	288.3	217.6	194.1
E	191.6	261.8	301.3	262.4	164.9	230.0

TENSION 2

Keyword: Tension, mass per volume.
Subject: The tensile strength,related to mass per volume and E-modulus.
Source: McLaughlin, E.C. A note on the strength of Jamaica grown bamboo. Wood and Fiber, vol 11 no 2 (1979), pp 86-91.

Results: The author has done tensile tests on specimens from Bambusa vulgaris, at 28 degree Celsius and 70% R.H., and with a straining speed of about 0.0007 per second. At first he has cut ten specimens of 300 mm length and a crosssection in the range 30 to 300 mm^2; the last in order to avoid size effects. With three-point bending tests the E-modulus has been determined. Next, the same specimens have been reduced to dumbbell-shaped specimens with a cross-section of 10 to 100 mm^2 in the narrow part. Tensile tests have been performed on these specimens, which are called "large samples".

From the fractured regions of these large samples, samples with a cross-section of 0.004 to 0.025 mm^2 were cut, and tensile tests on these so-called "fibers" has been carried out. From the ten samples only 8 with a failure in the middle were analysed. The original article gives full information on the test procedure.

The author calculated the following regression equations:
for the large samples: $\sigma = 0.76*rho + 2.25 * 0.001 * E - 131$
for the fibers: $\sigma = 0.25*rho + 21.4 * 0.001 * E - 443$
in which:
σ = the ultimate tensile stress in N/mm^2,
rho = the mass per volume of the bamboo in kg/m^3,
E = the Young's modulus in N/mm^2.

More data are as follows.

	large	fibers
sq.root of coeff. of determ.	0.841	0.847
E in 1000 N/mm^2 between	19-55	25-60
rho in kg/m^3 between	450-650	300-1500
strain at failure in %	1	1-3
ultim. tens. stress in N/mm^2	300-550	270-1100.

Example: the ultimate tensile stress for the mean values for large samples can be estimated as:

$$\sigma = 0.76 * 550 + 2.25 * 37 - 131 = 370 \text{ N/mm}^2.$$

TENSION 3

Keyword: Tension, mass per volume, site quality.
Subject: The ultimate tensile stress related to the mass per volume.
Source: Zhou Fangchun, Studies on physical and mechanical properties of bamboo wood. Journal of Nanjing Techn. Coll. of For. Prod., 1981 no 2 pp 1-32. In Chinese; the following is based on an English translation of the headings of the tables and the captions of the figures only. Cited here are tables 20 and 21.

Results: from table 20:

region	rho	σ	n
Yixing	612	196.1	293
Xiashu	596	180.2	128
Shimen	589	182.0	197
Damaoshan	584	174.0	421

in which:
rho = mass per volume in kg/m^3,
σ = ultimate tensile stress in N/mm^2,
n = number of tests.
Regression formula: $\sigma = 0.307*rho$; variation coefficient V is only 3%.

Results from table 20:

site class of forest	rho	σ	n
good	591	177.1	235
good/medium	597	181.0	122
medium	603	181.3	303
bad	602	194.9	239

with the same regression formula, with V = 4%.

TENSION 4

Keyword: Age, tension.
Subject: The ultimate tensile stress in Phyll.glauca cut at ages from 1 to 7 years.
Source: Lu Xiu-xin, Wang Ke-qing, Yi Xiang-chen, Liou Jin and He Ji-xun. A study on the physico-mechanical properties of culmwood of Phyll. glauca of Shandong. J. of bamboo res., Zhejiang For.Inst, Hangzhou, China, vol 4 no 2 (July 1985) pp 98-106.

Results: The authors have done tests on 54 culms from 4 different regions in the Shandong province; age classes from 1 to 7 years, and two culms per region and yearclass. Tests have been done on moisture content, mass per volume, tension, compression and bending, to determine the influences of the age at time of cutting and the region of origin. The number of tests was over 200 for each test and each region. The results dealing with the ultimate tensile stress have been determined on specimens with a total length of 250 mm, both ends in the form of a spoon, width 10 mm, and in the centre a length of 120 mm, width 1.5 mm. The thickness is equal to the wallthickness. Results:

region age	Sancha	Dajinkou	Dahuaya	Luchanya
1 year	305	232	275	218 N/mm²
2	313	265	269	261
3	321	284	281	273
4	343	302	242	327
5	328	293	281	279
6	322	283	275	280
7	322	280	272	

The polynomials are: (a = age in years)

region	tensile stress in N/mm²	st.dev.	best age
Sancha	$\sigma = 285 + 20.3*a - 0.22*a^2$	8.85 N/mm²	4.63 year
Dajinkou	$196 + 43.0*a - 4.54*a^2$	7.38	4.73
Dahuaya	$256 + 8.0*a - 0.99*a^2$	4.86	4.04
Luchanya	$155 + 68.4*a - 8.06*a^2$	25.2	4.25

TENSION 5

Keyword: Tension related to age.
Subject: The ultimate tensile stress related to the age.
Source: Zhou Fangchun, Studies on physical and mechanical properties of bamboo wood. Journal of Nanjing Techn. Coll. of For. Prod., 1981 no 2 pp 1-32. In Chinese; the following is based on an English translation of the headings of the tables and the captions of the figures only. Cited here is table 11 and figure 6.

Results: The ultimate tensile stress as function of the age of the bamboos at the time of cutting is in N/mm^2:

age class A	1	2	3	4	5	
age in years	1-2	3-4	5-6	7-8	9-10	n
district of origin						
Xiashu	151.96	189.00	178.89	199.60	181.74	128
Yixing	186.18	209.41	197.67	201.65	185.39	293
Shimen	163.92	191.22	194.82	184.46	169.98	197
Damaoshan	137.11	185.46	187.50	187.00	172.88	421
mean value	159.79	193.77	189.72	193.18	177.50	1039

A regression formula for the mean is:

$$\sigma = 132.76 + 38.29*A - 5.87*A^2.$$

The best age is 6.5 years.

TENSION 6

Keyword: Tension, position.
Subject: The ultimate tensile stress and the E-modulus, related to the position in the thickness of the culmwall.
Source: Duff, C.H. Bamboo and its structural use. The Eng. Soc. of China, session 1940-41, paper no. I.C.E. 1, Inst. of Civ. Eng., Shanghai 1941, pp 1-27. Cited here is figure 9.

Results: The author has determined the ultimate tensile stress and the E-modulus for airdry bamboo, across the thickness of the culmwall.
The next plot shows these two values in N/mm². In stead of E a value of E/500 has been plotted; consequently the vertical axis for E runs from 0 to 20000 N/mm². The horizontal axis represents the thickness of the culm-wall, left is the outside, right the inside. The numbers 0, 2, 4 and 6 are meaningless.

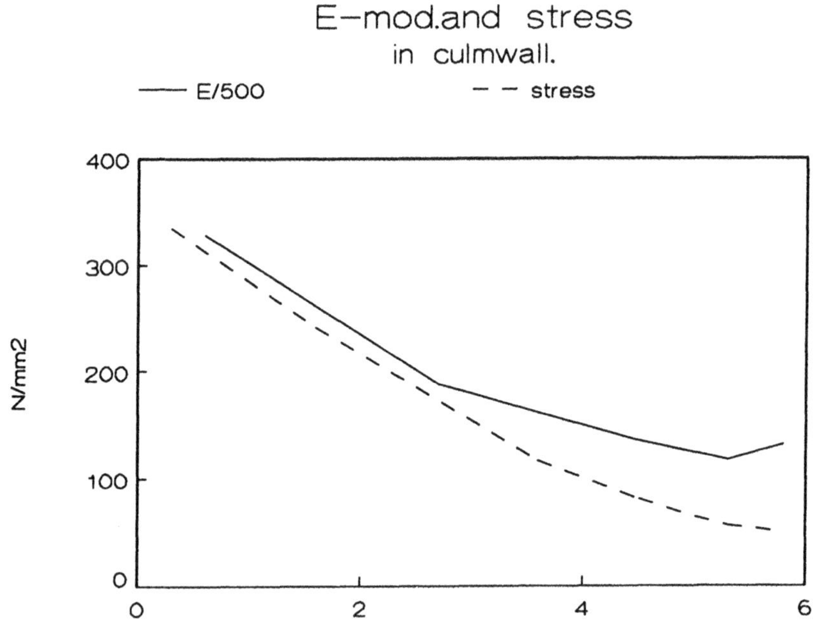

TENSION 7

Keyword: Tension, position.
Subject: Some data on tension related to the position in the culm.
Source: Atrops, J.L. Elastizitaet und Festigkeit von Bambusrohren (elasticity and strength of bambooculms). Der Bauingenieur (1969) vol 44 no 6 pp 220-225. In German.

Results: Atrops has done tests on tension. He reports about some data, as follows.

1. The mean ultimate tensile stress σ is:
   in the outer layer of the culm-wall: 287 N/mm^2,
   in the inner layer: 151 N/mm^2.
   Number of tests 30 each;cross-section of specimen 1 mm x 10 mm.
2. The mean ultimate tensile stress, tested on specimens as thick as the culm-wall, is 210 N/mm^2, with a variation coefficient V of 10%. At the top of the culm this stress is 12% less than at the bottom.
3. E-modulus 17900 to 24100 N/mm^2.

TENSION 8

Keyword: Tension, seasoning.
Subject: The ultimate tensile stress and the E-modulus as function of the number of days after cutting.
Source: Cox and Geymayer. Expedient reinforcement for concrete for use in S.E.Asia. 1969. Cited here is table 5.

Results:

j	-	MC	σ	node		internode	
				E	E/σ	E	E/σ
3 to 4	m	95.16	107.28	16780	141.7	16290	157.6
	s	52.84	33.28	4880	38.6	4390	50.6
	n	17	18	6	6	9	9
20	m	16.8	117.8	20810	169.6	19405	178.5
	s	2.0	27.1	4970	28.8	1325	42.0
	n	11	11	4	4	7	7
142	m	10.1	91.6	19180	229.3	20065	238.7
	s	0.2	30.5	3030	84.2	3675	86.9
	n	5	5	5	5	5	5

in which:
j = number of days after cutting,
m = mean, s = standard deviation, n = number of tests,
MC = moisture content in %,
σ = ultimate tensile stress in N/mm²,
E = E-modulus.
Test method: whole culms, Arundinaria tecta, fixed with Chinese pullers, speed of loading 0.12 N/mm²/s. From the specimens 76.5 precent failed at nodes; however there is no significant difference between the stress at a node and at an internode:

	σ at node	σ at internode
m	107.5	106.5 N/mm²
s	32	43
n	14	4

TENSION 9

Keyword: Poisson's ratio.
Subject: The Poisson's ratio in tension tests.
Source: Cox and Geymayer. Expedient reinforcement for concrete for use in S.E. Asia. 1969. Cited here is table 6.

Results: The authors have determined the Poisson's ratio in tension tests on Arundinaria tecta, with strain gauges.

j	-	MC	Pr
4	m	97.0	0.307
	s	68.6	0.076
	n	3	3
20	m	16.9	0.321
	s	2.2	0.047
	n	8	8
-	m		0.317
	s		0.052
	n		11

in which:
j = number of days after cutting,
m = mean, s = standard deviation, n = number of tests,
MC = moisture content in %,
Pr = Poisson's ratio.

10. Torsion

TORSION 1

Keyword: Torsion.
Subject: Torsion tests on split bamboos.
Source: Ueda, K. Mechanical properties of Moso bamboo (Phyll.pub.). In: Res. Bull. Coll. Exp. For. Hokkaido Univ., vol 37 no 3 (1980), pp 817-836.

Results: The author has done tests on bending, compression and torsion on 51 specimens of Phyll.pub. The specimens are about 10 mm thick and 10 mm wide. The tests on torsion have been reported as follows.

no. of specimen	rho kg/m^3	t/b	G N/mm^2
03	730	1.067	1050
24	750	1.205	1180
25	750	1.204	1100
35	730	1.175	1080
36	720	1.100	1060
39	720	1.206	1020
41	730	1.273	1130
mean			1090

in which:
t/b = the thickness/width ratio,
G = the modulus of rigidity in N/mm^2.

11. Relations between properties

RELATIONS BETWEEN PROPERTIES 1

Keyword: Anatomy, mechanical properties.
Subject: The quantity and dimensions of fibers and vessels in Bambusa blumeana and Gigantochloa levis, related to moisture content, mass per volume, compression and bending.
Source: Espiloy, Z.B., A.B. Ella and A.R. Floresca. Physico-mechanical properties and anatomical structure relationships of two erect bamboo species. The Philippine Lumberman, April 1986, pp 25, 26, 27, 35, and May 1986 pp 32, 33, 35.

Results: The authors did tests on five culms each of three years old of each species. From bottom, middle and top segments of 8 internodes were cut. The dimensions of fibers and vessels were determined after maceration in 60% glacial acetic acid and 30% hydrogen peroxide, and then shaking in 50-75% ethyl alcohol. Dimensions were taken under a microprojector with 40X magnification. The number of vascular bundles has been determined with 20X magnification, and 20 observations per sample. Results:

species	Bambusa blumeana				Gigantochloa levis			
position	bott.	midd.	top	mean	bott.	midd.	top	mean
n/mm^2	1.74	3.12	3.80	2.89	1.11	1.50	2.06	1.56
fiberdimensions in mm								
L	2.50	2.55	2.63	2.56	1.88	1.52	2.05	1.82
D	0.0148	0.0152	0.0142	0.0147	0.0165	0.0155	0.0159	0.0160
LD	0.0046	0.0032	0.0034	0.0037	0.0056	0.0036	0.0065	0.0052
CWT	0.0052	0.0061	0.0054	0.0056	0.0060	0.0058	0.0059	0.0059
vesseldimensions in mm								
L	0.8228	0.7068	0.8294	0.7863	1.1305	0.6776	0.5957	0.8013
D	0.1862	0.1366	0.1736	0.1655	0.2310	0.2308	0.1989	0.2202

in which: n/mm^2 = number of fibrovascular bundles per mm^2, L = length, D = diameter, LD = lumendiameter, CWT = cellwall-thickness.

A statistical analysis gives the means squares as follows.

source of var.	bottom/middle/top		species		CV(X)
fibrovasc.bundles	179.681	**	837.801	**	56.55
fiber length	0.238	ns	4.019	**	14.49
fiber diameter	3.229	ns	43.471	**	26.85
lumen diameter	19.769	**	17.170	**	38.09
cellwallthickness	0.420	**	6.914	**	5.49
vessel length	1496.542	**	4945.081	**	20.64
vessel diameter	43.972	ns	1314.614	**	24.33

in which ** means significant at 99% level of probability, and ns means not significant.
The original document gives full tables with the correlation coefficients between the number of vascular bundles per mm^2, the fiber length, the fiber diameter, the lumen diameter, the cellwall thickness, the vessel length and the vessel diameter, with the moisture content, the mass per volume, the ultimate stresses in compression and bending, the elastic stress in bending, and the E-modulus in bending, for both species.

Significant at 95% level are only the following items.

For B.blumeana:
-the number of vascular bundles per mm^2, with elastic bending stress, and with E-modulus,
-the fiber length with elastic bending stress,
-the lumen diameter negatively with the ultimate compression stress.

For G.levis:
-the number of vascular bundles per mm^2, with the E-modulus,
-the fiber length with the elastic bending stress.

Finally, the following correlation coefficients have been calculated (only those, significant at 95% are given here).

characteristics of specimen	outside diameter	length/span ratio	culmwall thickness
B.blumeana			
compression			
σ'	-	-	-
rho	-	-0.436	-0.436
bending			
σ.el	-	-	0.617
σ ult.	-	-	0.670
E	-	-	-
rho	-	-	-
G.levis			
compression			
σ'	-0.448	-0.668	-0.668
rho	-0.434	-0.441	-0.441
bending			
σ.el	-	-	-
σ ult.	-	-	-
E	-	-	-0.795
rho	-0.789	-	-0.615

RELATIONS BETWEEN PROPERTIES 2

Keyword: Fibre, mechanical properties.
Subject: The lengths of fibres in various nodes of four species related to bending, compresion and tension.
Source: Widjaja, E. and Z. Risyad. Anatomical properties of some bamboos utilized in Indonesia. Pp 244-249 in Bamboo research in Asia. Proc. workshop Singapore May 1980. Editors Lessard, G. and Chouinard, A. Ottawa, ISBN: 0-88936-267-X.

Results:

species	node	fibre length	E	σb	σ'	σt
D.giganteus	1st	19.16	17210	183	60.3	184
	3rd	18.04	12246	176	62.0	195
	5th	19.16	14791	183	64.0	188
	7th	19.0	13035	288	64.6	197
	mean	18.84	14320	182	62.7	191
D.asper	1st	19.28	12207	164	63.9	215
	3rd	19.60	14960	174	59.2	204
	5th	21.16	12954	160	62.2	222
	7th	20.10	12397	158	56.6	210
	mean	20.03	13129	164	60.5	213
G.robusta	1st	19.66	9421	138	53.3	197
	3rd	15	9237	129	51.0	177
	5th	19.66	10922	140	51.1	185
	7th	18.16	9738	135	53.0	207
	mean	18.12	9829	136	52.1	191
B.vulg.striata	1st	19.92	6065	107	48.4	139
	3rd	20.24	7193	112	44.3	120
	5th	18.2	8830	111	47.6	135
	7th	17.2	8394	129	41.7	135
	mean	18.92	7621	115	45.5	132

in which fibre length is in µm,
E = E-modulus, σb = ultimate bending stress, σ' = ultimate compression stress and σt = ultimate tensile stress, all in N/mm^2.
Linear regression gives the next formulae (L = fibre length)

Formula:	corr. coeff.	F
E = 3057 + 417*L	0.3	8.48
σb= 148 + 0.033*L	0.0001	0.00096
σ'=-4.27 + 3.13*L	0.03	11.09
σt= 160 + 1.35*L	0.1	0.99

RELATIONS BETWEEN PROPERTIES 3

Keyword: Site quality, mechanical properties.

Subject: The diameter, mass per volume, compression, tension and bending stresses in Phyll.glauca cut from four different regions.

Source: Lu Xiu-xin, Wang Ke-qing, Yi Xiang-chen, Liou Jin and He Ji-xun. A study on the physico-mechanical properties of culmwood of Phyll.glauca of Shandong. J. of bamboo res., Zhejiang For.Inst., Hangzhou, China, vol 4 no 2 (July 1985) pp 98-106.

Results: The authors have done tests on 54 culms of 4 different regions in the Shandong province; age classes from 1 to 7 years, and two culms per region and yearclass. Tests have been done on moisture content, mass per volume, tension, compression and bending, to determine the influences of the age at time of cutting and the region of origin. The number of tests was over 200 for each test and each region. The results have been compared with the region of origin of the bamboos:

Region	Luchanya	Sancha	Dahuaya	Dajinkou
situation	middle of a hill	cold mountain area, along a river.	ocean-climate, flat land in branch of river	.river flat land near river
oriented	S-N	E-W		S-N.
altitude	400 m + sea	800		
water-supply	extremely rare	reasonable	reasonable (mech.pump)	good (3 wells)
RH (1)	54%	55	59	62
org.subst.	3.02%	3.67	1.58	9.40
N	0.166%	0.17	0.11	0.57
P	0.53%	0.208	0.23	0.79
K	62.3 ppm	196.1	120.35	305.6
Properties:				
diameter	46.6 mm	47.0	49.4	48.5 mm
rho	717	704	680	679 kg/m³
compr.str.	82.7	73.2	71.0	65.8N/mm²
tens.str.	296	329	277	289 N/mm²
bend.str.	186	174	162	153 N/mm²

(1): relative humidity during bamboo-shoots-period.
rho = mass per volume.
Remarkably the left column shows the best properties,while the quality of the site is not that good. In the right column, the site looks best; the properties however are the worst. The good relation between mass per volume and mechanical properties can easily been seen.

RELATIONS BETWEEN PROPERTIES 4

Keyword: Site quality, vessel, mass per volume, compression and tension.
Subject: The effects of the quality of the forest site on the said physical and mechanical properties.
Source: Zhou Fangchun, Studies on physical and mechanical properties of bamboo wood. Journal of Nanjing Techn. Coll. of For. Prod., 1981 no 2 pp 1-32. In Chinese; the following is based on an English translation of the headings of the tables and the captions of the figures only. Cited here is table 20.

Results:

site class	dbh	rho	vasc.	compression no	compression σ'	tension no	tension σ
	mm	kg/m³	n/mm²		N/mm²		N/mm²
good	125	591	1.87	288	61.8	235	177
go/med	105	597	1.96	140	64.7	122	181
medium	98	603	2.22	376	63.2	303	181
bad	81	602	2.40	299	65.8	239	195

in which:
dbh = diameter at breast height in mm,
rho = mass per volume in kg/m^3,
vasc. = number of vascular bundles per mm^2,
no = number of tests,
σ' = ultimate compression stress in N/mm^2,
σ = ultimate tension stress in N/mm^2,
go/med = good to medium.

Note by the present author:
From the results published by this researcher we might draw the conclusion as follows. A better site class results in an increase of the diameter, and in a decrease of the mass per volume, the number of vascular bundles and the compression and tension strength. The same conclusion occurs from figure 8 in the original article (not cited here).

RELATIONS BETWEEN PROPERTIES 5

Keyword: Age, mechanical properties.
Subject: The mass per volume, and the ultimate stresses in bending, compression, tension, shear, and cleavage, related to the age at the time of cutting.
Source: Li Yunlien and Li Yezhen. Physico-mechanical properties of culmwood of Phyll.pub. produced in Guizhou. Bamboo research, Nanjing techn. coll. of for. prod. (1983), no 1, pp 52-74.

Results: In October 1980 90 culms have been collected from 21 fields. All culms were normal, with a diameter of 100 mm or more and a length of 15 m. Tests were done on specimens from the lowest 6 m. The thickness of all specimens was their wallthickness.

Other sizes were:
-for the mass per volume and compression tests: 20 mm to 20 mm,
-for bending: 300 mm long and 10 mm wide, 3-point bending test,
-for tension: 260 mm long, and in the centre a length of 60 mm with a width of 4 mm,
-for shear 50 mm high and 18 mm wide with a notch on one side of 12 mm wide and 30 mm high,
-for cleavage 50 mm long and 20 mm wide with a < of 25 mm deep.
The original paper gives full statistical data, like number of tests (about 60 or 70 per test), mean value, standard deviation, standard error, variation coefficient and accuracy index. The present author consider these data (how interesting they are) as too specific to be quoted here. A summary is:

age-class	I	II	III	IV	V	
years	1-2	3-4	5-6	7-8	9-10	
rho	542	583	620	641	630	kg/m^3
tang.bend.σ	118	129	141	147	144	N/mm^2
rad.bend.σ	96	104	113	119	114	N/mm^2
compr. σ'	45	49	53	54	52	N/mm^2
tension σ	106	161	186	188	187	N/mm^2
shear σv	8.3	10.1	11.7	12.1	11.6	N/mm^2
cleavage	1.7	1.9	2.1	2.3	2.2	N/mm^2

The original document gives also regression formula, e.g. for the mass per volume:
$rho = 473.5 + 74.7*a - 8.6*a^2$ with correlation factor $r = .999$;
a = ageclass. From such formulae the best age for the bamboo to be cut can be calculated; in this particular case this age is 8 years. The original paper gives also data on shrinkage and swelling.

RELATIONS BETWEEN PROPERTIES 6

Keyword: Mechanical properties, position.
Subject: The density of vascular bundles, the mass per volume, and the ultimate stresses in bending, compression, tension, shear, and cleavage, related to the position in the culm.
Source: Li Yunlien and Li Yezhen. Physico-mechanical properties of culmwood of Phyll.pub. produced in Guizhou. Bamboo research, Nanjing techn. coll. of for.prod. (1983), no 1, pp 52-74.

Results: In October 1980 90 culms have been collected from 21 fields. All culms were normal, with a diameter of 100 mm or more and a length of 15 m. Tests were done on specimens from the lowest 6 m. The thickness of all specimens was their wallthickness. Other sizes were:
-for the mass per volume and compression tests: 20 mm to 20 mm,
-for bending: 300 mm long and 10 mm wide, 3-point bending test,
-for tension: 260 mm long, and in the centre a length of 60 mm with a width of 4 mm,
-for shear 50 mm high and 18 mm wide with a notch on one side of 12 mm wide and 30 mm high,
-for cleavage 50 mm long and 20 mm wide with a < of 25 mm deep.

The lowest 3 m were divided into 3 pieces of 2 m each. Results of tests are as follows.

position	1-2 m	3-4 m	5-6 m	
no. vasc. b.	1.58	2.39	2.67	n/mm^2
rho	554	635	660	kg/m^3
tang.bend.σ	127	147	149	N/mm^2
rad. bend.σ	116	120	125	N/mm^2
compr. σ'	41.6	53.2	55.9	N/mm^2
tension σ	139	185	202	N/mm^2
shear σv	11.2	12.0	13.2	N/mm^2
cleavage	2.2	2.3	2.4	N/mm^2

in which: no.vasc.b. = number of vascular bundles per mm^2,
rho = mass per volume,
and the remaining data concern ultimate stresses, in the case of tangential and radial bending, compression, tension, shear and cleavage resp.
The increase of vascular bundles from bottom to top can be seen easily, as well as the increase of the mechanical properties. This last is due to the fibres, which are mixed up with the vascular bundles (the vascular bundles consist out of vessels and fibres).
The original paper gives also full statistical data on this relation.

RELATIONS BETWEEN PROPERTIES 7

Keyword: Moisture content, mass per volume, mechanical properties, position.

Subject: The moisture content, mass per volume, compression and bending in Bambusa blumeana and Gigantochloa levis, related to the position in the culm.

Source: Espiloy, Z.B., A.B. Ella and A.R. Floresca. Physico-mechanical properties and anatomical structure relationships of two erect bamboo species. The Philippine Lumberman, April 1986, pp 25, 26, 27, 35, and May 1986 pp 32, 33, 35.

Results: The authors did tests on five culms each of three years old of each species. From bottom, middle and top segments of 8 internodes were cut. From these segments the next specimens have been prepared.

-moisture content and mass per volume: 25.4 mm x 25.4 mm x wallthickness. The volumes have been determined with the liquid displacement method in water, and then ovendried.

-bending tests have been done on full green culms, with a free span equal to 14 times the diameter; three point bending test.

-compression tests were done on specimens with a height of 10 times the wallthickness. Results:

species	Bambusa blumeana				Gigantochloa levis			
position	bott.	midd.	top	mean	bott.	midd.	top	mean
MC in %	195	114	99	136	143	115	94	117
rho in kg/m³	388	537	585	503	474	539	610	541
σ' in N/mm²,								
nodal	34.7	37.2	37.3	36.4	37.7	41.9	44.3	41.3
internodal	35.9	39.5	39.5	38.3	38.7	41.7	43.4	41.3
bending:								
σ.el in N/mm²	29.5	20.2	18.2	22.6	17.6	14.9	18.6	17.0
σ ult. in N/mm²	43.2	28.4	24.7	32.1	25.4	19.6	26.1	23.7
E in 1000N/mm²	8.9	8.8	9.2	9.0	8.9	10.4	11.0	10.1

in which:
MC = moisture content in %,
rho = mass per volume, ratio dry weight to green volume, kg/m^3,
σ' = ultimate compression stress,
σ.el =bending stress at limit of elasticity (proportional limit)
σ ult = ultimate bending stress = modulus of rupture,
E = modulus of elasticity. All compression and bending on green bamboo.

Statistical analysis:

Source of var.	bottom/middle/top		species		CV(X) %
MC	51129.90	*	13038.32	ns	63.88
rho	0.3321	*	0.0300	ns	29.60
σ'	5443.019	**	24040.017	**	3.97
σ.el	10884.936	ns	24741.482	ns	48.67
σ ult.	32240.700	ns	55126.533	ns	54.35
E	300.237	ns	909.900	ns	17.26

in which * = significant at 95% level, ** = idem at 99% and ns = not significant.
The correlation between fibres and vessels, and mechanical properties, has also been reported. The original report deals also with shrinkage.

Keyword: Moisture content related to mechanical properties.
Subject: The effect of water absorption on properties.
Source: Godbole, V.S. and S.C. Lakkad. Effect of water absorption on the mechanical properties of bamboo. J. of Mat. Sc. letters, vol 5, 1986, pp 303-304.
Results: The author has prepared bamboo samples in different ways: -dry; coated in epoxy and soaked in distilled water for 144 hours, resulting in a MC of 7.5%; boiled in distilled water for 2 hours, resulting in a MC of 38.2%; and soaked in distilled water for 144 hours, MC of 81.2%.
Results are as follows.

M.C. in %	E in N/mm^2	σ in N/mm^2	σ' in N/mm^2	σv in N/mm^2
0	22500	238	63.2	9.99
7.5	21500	201	54.8	9.30
38.2	15500	175	35.8	6.36
81.2	11760	150	43.3	6.98

A simple regression analysis shows the next formulae.

E-modulus: $E = 22786 - 233*MC + 1.19*MC^2$, with R =.998,
compression: $\sigma = 228 - 2.01*MC + 0.013*MC^2$, .966,
tension: $\sigma' = 62.6 - 0.98*MC + 0.008*MC^2$, .999,
shear: $\sigma v = 10.13 - 0.15*MC + 0.001*MC^2$, .996.

A plot is:

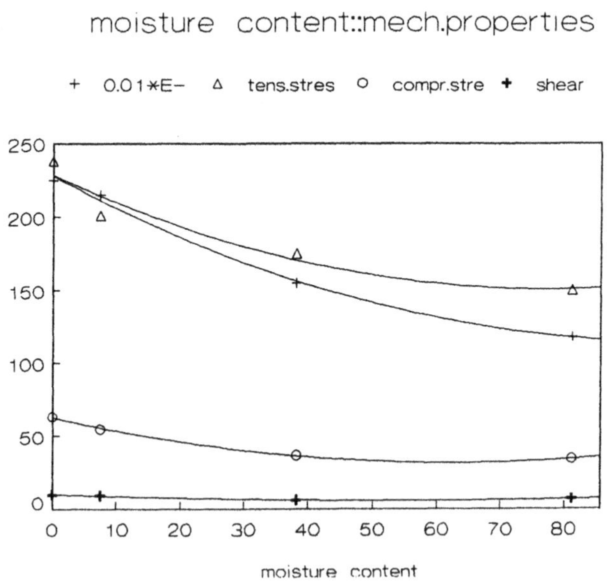

RELATIONS BETWEEN PROPERTIES 9

Keyword: Site quality, mass per volume, mechanical properties.
Subject: Mass per volume, MC, bending and compresion for Dendr. Strictus from different regions.
Source: Sekhar, A.C. and A.S. Gulati. A note on the physical and mechanical properties of Dendrocalamus strictus from different localities. Van Vigyan (J. of the Soc. of Indian For.), vol 11 nos 3/4 (Sept.-Dec. 1973) pp 17-22.

Results: Seven consignments from the said species of different localities, consisting of 10 culms each of about 3.3 m long, have been tested. Specimens for bending were prepared with an internode in the centre; specimens for compression were node and internode, 50% each. The regions and the local names were:

A. Mohand range, Dehra Dun, U.P., Bamboo.
B. North Chak Range Gorakhpur U.P., Kathibansi.
C. Luxectipet Range Mancherial Div., Adilabad A.P., Bamboo.
D. Nagar Karnoo Mananur A.P., Solid bamboo Gathibans.
E. Chintur Range Bhadrachelum A.P., Sadanam Veduru.
F. Kagaj Nagar A.P., Vedupu Bonga, Kank.
G. Sukna Range W.B., Lathi Bans.

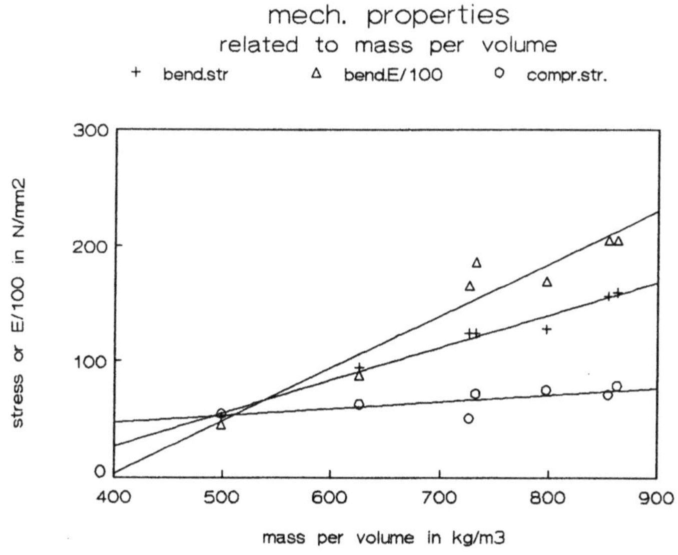

Test results are as follows.

region (as before)	A	B	C	D	E	F	G
age in years	2-4	4	3	2-4	2	3	3
nodes (note a)	-	6	5	4	6	6	5
D in mm, mean	78.8	70.5	43.8	31.6	44.2	37.4	43.4
max.	90.3	79.2	55.9	48.9	57.9	49.9	51.3
min.	66.0	62.9	29.5	17.5	31.0	28.7	34.4
d in mm, mean	53.3	33.1	26.9	12.0	26.9	20.1	17.9
max.	65.9	46.7	39.3	20.6	38.7	31.5	29.2
min.	30.4	13.4	15.5	0.00	17.8	14.3	0.42
t in mm, mean	12.7	18.7	8.3	9.8	9.3	8.5	12.9
max.	21.3	26.7	13.3	18.2	13.4	15.5	20.0
min.	7.1	13.7	5.0	6.0	5.2	5.0	6.7
green							
rho in kg/m³	428	538	696	639	710	776	-
MC	154	92	24	34	29	24	-
bending, σ.el	21.6	42.5	36.7	55.9	36.1	47.8	-
σ	42.8	75.3	69.8	98.5	64.9	78.5	-
E	4360	6460	13440	18310	12210	13900	-
compr., σ'	26.1	34.4	31.9	45.1	34.3	43.3	-
air-dry							
rho in kg/m³	499	626	732	854	797	862	726
MC	8	10	8	11	8	9	13
bending, σ.el	32.1	66.5	79.7	108.9	93.3	118.6	75.7
σ	51.9	94.1	123.9	156.8	127.8	159.7	124.2
E	4570	8760	18600	20500	17020	20530	16560
compr. σ'	54.8	62.9	71.8	71.4	74.9	78.8	50.7

note a=average number of nodes between group level and 1 m ring. In this table is: D = outside diameter, d =inside dia. and t = wallthickness, all in mm. MC = moisture content in %.

The stresses: σ.el = stress at limit of elasticity,
σ = ultimate bending stress,
E = E-modulus,
σ' = ultimate compression stress, all in N/mm².

A simple linear regression analysis shows the next relations between the mechanical properties and the mass per volume:
Ultimate bending stress $\sigma = -86 + 0.28 * rho$ with $R = 0.99$,
E-modulus in bending $E = -17800 + 45 * rho$ $R = 0.96$,
Ultimate compression $\sigma' = 23.8 + 0.058 * rho$ $R = 0.72$.
Stresses and E in N/mm², and rho in kg/m³.
R = square root of coeff. of determination.
A plot for these regressions is shown on the previous page (ultimate bending stress, 0.01 * E in bending, and ultimate compression stress resp.)

List of symbols

Symbol	*Unit*	*Definition*
d	mm	diameter (sometimes: D = outside diameter, and d = inside diameter.)
E	N/mm^2	modulus of elasticity, Young's modulus.
F	N	external force
H	-	height of the culm
j	-	number of days
m	-	mean of a sample
MC	%	moisture content
n	-	number of specimens
R	-	square root of coefficient of determination
rho	kg/m^3	mass per volume, (density)
s		standard deviation of a sample
σ	N/mm^2	ultimate stress in bending or tension
σ'	N/mm^2	ultimate stress in compression
σ.el	N/mm^2	stress at limit of elasticity
σv	N/mm^2	ultimate stress in shear
V	%	coefficient of variation = (s/m)*100%
t	mm	wallthickness
w	mm	deformation

Author index

Keyword index

The next page presents an index on keywords as well, but with an emphasis on the relations between properties. If e.g. some reader likes to find an article on *compression* in relation with the *age* of the bamboo at the time of cutting, and the *mass per volume*, then this article can be traced with the keywords "age compress. mass" on the next page. (some keywords are given in an abbreviated form, and all are given in alphabetical order). The result is page 84.

Keyword index (relations between properties)

FORESTRY SCIENCES

1. P. Baas (ed.): *New Perspectives in Wood Anatomy.* Published on the Occasion of the 50th Anniversary of the International Association of Wood Anatomists. 1982 ISBN 90-247-2526-7
2. C.F.L. Prins (ed.): *Production, Marketing and Use of Finger-Jointed Sawnwood.* Proceedings of an International Seminar Organized by the Timber Committee of the UNECE (Halmar, Norway, 1980). 1982 ISBN 90-247-2569-0
3. R.A.A. Oldeman (ed.): *Tropical Hardwood Utilization.* Practice and Prospects. 1982 ISBN 90-247-2581-X
4. P. den Ouden (in collaboration with B.K. Boom): *Manual of Cultivated Conifers.* Hardy in the Cold- and Warm-Temperate Zone. 3rd ed., 1982 ISBN Hb 90-247-2148-2; Pb 90-247-2644-1
5. J.M. Bonga and D.J. Durzan (eds.): *Tissue Culture in Forestry.* 1982 ISBN 90-247-2660-3
6. T. Satoo: *Forest Biomass.* Rev. ed. by H.A.I. Madgwick. 1982 ISBN 90-247-2710-3
7. Tran Van Nao (ed.): *Forest Fire Prevention and Control.* Proceedings of an International Seminar Organized by the Timber Committee of the UNECE (Warsaw, Poland, 1981). 1982 ISBN 90-247-3050-3
8. J.J. Douglas: *A Re-Appraisal of Forestry Development in Developing Countries.* 1983 ISBN 90-247-2830-4
9. J.C. Gordon and C.T. Wheeler (eds.): *Biological Nitrogen Fixation in Forest Ecosystems.* Foundations and Applications. 1983 ISBN 90-247-2849-5
10. M. Németh: *Virus, Mycoplasma and Rickettsia Diseases of Fruit Trees.* Rev. (English) ed., 1986 ISBN 90-247-2868-1
11. M.L. Duryea and T.D. Landis (eds.): *Forest Nursery Manual.* Production of Bareroot Seedlings. 1984; 2nd printing 1987 ISBN Hb 90-247-2913-0; Pb 90-247-2914-9
12. F.C. Hummel: *Forest Policy.* A Contribution to Resource Development. 1984 ISBN 90-247-2883-5
13. P.D. Manion (ed.): *Scleroderris Canker of Conifers.* Proceedings of an International Symposium on Scleroderris Canker of Conifers (Syracuse, USA, 1983). 1984 ISBN 90-247-2912-2
14. M.L. Duryea and G.N. Brown (eds.): *Seedling Physiology and Reforestation Success.* Proceedings of the Physiology Working Group, Technical Session, Society of American Foresters National Convention (Portland, Oregon, USA, 1983). 1984 ISBN 90-247-2949-1
15. K.A.G. Staaf and N.A. Wiksten (eds.): *Tree Harvesting Techniques.* 1984 ISBN 90-247-2994-7
16. J.D. Boyd: *Biophysical Control of Microfibril Orientation in Plant Cell Walls.* Aquatic and Terrestrial Plants Including Trees. 1985 ISBN 90-247-3101-1
17. W.P.K. Findlay (ed.): *Preservation of Timber in the Tropics.* 1985 ISBN 90-247-3112-7
18. I. Samset: *Winch and Cable Systems.* 1985 ISBN 90-247-3205-0
19. R.A. Leary: *Interaction Theory in Forest Ecology and Management.* 1985 ISBN 90-247-3220-4
20. S.P. Gessel (ed.): *Forest Site and Productivity.* 1986 ISBN 90-247-3284-0

FORESTRY SCIENCES

21. T.C. Hennessey, P.M. Dougherty, S.V. Kossuth and J.D. Johnson (eds.): *Stress Physiology and Forest Productivity*. Proceedings of the Physiology Working Group, Technical Session, Society of American Foresters National Convention (Fort Collins, Colorado, USA, 1985). 1986 ISBN 90-247-3359-6
22. K.R. Shepherd: *Plantation Silviculture*. 1986 ISBN 90-247-3379-0
23. S. Sohlberg and V.E. Sokolov (eds.): *Practical Application of Remote Sensing in Forestry*. Proceedings of a Seminar on the Practical Application of Remote Sensing in Forestry (Jönköping, Sweden, 1985). 1986 ISBN 90-247-3392-8
24. J.M. Bonga and D.J. Durzan (eds.): *Cell and Tissue Culure in Forestry*. Volume 1: General Principles and Biotechnology. 1987 ISBN 90-247-3430-4
25. J.M. Bonga and D.J. Durzan (eds.): *Cell and Tissue Culure in Forestry*. Volume 2: Specific Principles and Methods: Growth and Development. 1987 ISBN 90-247-3431-2
26. J.M. Bonga and D.J. Durzan (eds.): *Cell and Tissue Culure in Forestry*. Volume 3: Case Histories: Gymnosperms, Angiosperms and Palms. 1987 ISBN 90-247-3432-0
    Set ISBN (Volumes 24-26) 90-247-3433-9
27. E.G. Richards (ed.): *Forestry and the Forest Industries: Past and Future*. Major Developments in the Forest and Forest Industries Sector Since 1947 in Europe, the USSR and North America. In Commemoration of the 40th Anniversary of the Timber Committee of the UNECE. 1987 ISBN 90-247-3592-0
28. S.V. Kossuth and S.D. Ross (eds.): *Hormonal Control of Tree Growth*. Proceedings of the Physiology Working Group, Technical Session, Society of American Foresters National Convention (Birmingham, Alabama, USA, 1986). 1987 ISBN 90-247-3621-8
29. U. Sundberg and C.R. Silversides: *Operational Efficiency in Forestry*. Vol. 1: Analysis. 1988 ISBN 90-247-3683-8
30. M.R. Ahuja (ed.): *Somatic Cell Genetics of Woody Plants*. Proceedings of the IUFRO Working Party S2.04-07 Somatic Cell Genetics (Grosshansdorf, Germany, 1987). 1988. ISBN 90-247-3728-1
31. P.K.R. Nair (ed.): *Agroforestry Systems in the Tropics*. 1989 ISBN 90-247-3790-7
32. C.R. Silversides and U. Sundberg: *Operational Efficiency in Forestry*. Vol. 2: Practice. 1989 ISBN 0-7923-0063-7
    Set ISBN (Volumes 29 and 32) 90-247-3684-6
33. T.L. White and G.R. Hodge (eds.): *Predicting Breeding Values with Applications in Forest Tree Improvement*. 1989 ISBN 0-7923-0460-8
34. H.J. Welch: *The Conifer Manual*. Volume 1. 1991 ISBN 0-7923-0616-3
35. P.K.R. Nair, H.L. Gholz, M.L. Duryea (eds.): *Agroforestry Education and Training*. Present and Future. 1990 ISBN 0-7923-0864-6
36. M.L. Duryea and P.M. Dougherty (eds.): *Forest Regeneration Manual*. 1991 ISBN 0-7923-0960-X
37. J.J.A. Janssen: *Mechanical Properties of Bamboo*. 1991 ISBN 0-7923-1260-0

KLUWER ACADEMIC PUBLISHERS – DORDRECHT / BOSTON / LONDON

9 780792 312604

GPSR Compliance

The European Union's (EU) General Product Safety Regulation (GPSR) is a set of rules that requires consumer products to be safe and our obligations to ensure this.

If you have any concerns about our products, you can contact us on ProductSafety@springernature.com

In case Publisher is established outside the EU, the EU authorized representative is:

Springer Nature Customer Service Center GmbH
Europaplatz 3
69115 Heidelberg, Germany

Batch number: 10584448

Printed by Printforce, the Netherlands